SWARM TROOPERS

HOW SMALL DRONES WILL CONQUER THE WORLD

BY: DAVID HAMBLING
Swarm-Troopers.com

623.746

Publishing services provided by Archangel Ink

ISBN: 1-942761-74-0
ISBN-13: 978-1-942761-74-7

Special thanks to Philip Hemplow, Tim Radford, Jürgen Altmann, Eugene Tan, Mark Venguerov and Paul Edgington for helpful comments and suggestons on the manuscript.

Visit the website Swarm-Troopers.Com for a drone picture gallery, regular updates and news on the progress of small drone technology, and special offers.

TABLE OF CONTENTS

INTRODUCTION

The Farnborough International Airshow is a big event in the aviation world's calendar. Every two years the whole industry gets together for a week to make deals, scope out the competition, show off their latest wares and talk shop. The highlight of Farnborough 2014 was supposed to be the public debut of the F-35 Lighting II, a stealth warplane from Lockheed Martin.

Unfortunately, engine problems left the F-35 fleet grounded. The only one at Farnborough was a display model for people to sit in and have their picture taken.

Airpower is so dominant in modern wars that in a sense the F-35 represents the future of warfare, the cutting edge of the world's most formidable fighting machine. The US Air Force, Navy and Marine Corps plan on buying about two and a half thousand F-35s between them, at between $100 million and $200 million per plane. The exact price is fiercely debated, as we shall see. The aircraft's no-show was symptomatic of the many problems experienced during development, but the largest aircraft purchase in US history has gone too far to stop now. Allies including the UK, Italy, Israel, South Korea and Turkey have already placed their orders.

While the F-35 was absent, a much smaller aircraft did make its debut at Farnborough. The Micro Drone 2.0 is a palm-sized flying toy with four rotors. A salesman was showing how well it flew, automatically righting itself after being tossed into the air. He was doing a brisk trade, selling boxed drones over the counter at $85 apiece. What made the Micro Drone so appealing was a video camera that turned it from a toy into a tool.

"Everybody wants one because they're such great gadgets," the salesman told me. "And because you can actually do practical things like check the guttering, you can justify buying one to your wife."

The tiny Micro Drone did not seem to have much in common with the thundering jets tearing up the sky outside. However, it was similar to military quadrotors on display, slightly bigger, tougher, and more expensive versions of the same camera-carrying design. One, the AV Sparrow, was little bigger than the Micro Drone, others weighed a few pounds. Their fixed-wing counterparts, looking like radio-controlled models with a four-foot wingspan, have become one of the soldier's most trusted tools, as we will see in Chapter 3. Moving up in scale there are drones like the Puma, Scan Eagle, and Shadow, which resemble light aircraft. The largest, Shadow, has a wingspan of fourteen feet.

Stepping outdoors at Farnborough you could find the next size up, the Predator and the Reaper. These are armed drones the size of manned planes that spy and carry out strikes in Afghanistan, Pakistan, and elsewhere. These are still small fry in the aviation world. The F-35 is six times the weight of the Reaper, as well as being faster, stealthier, more agile, and able to carry a much greater load of weapons. At a fundamental level though, both are warplanes doing the same job, although in one case the pilot stays on the ground.

A souped-up Micro Drone could carry out a military mission, scouting enemy positions from a distance. Small drones can carry a lethal explosive warhead; US Special Forces have used them in Afghanistan against insurgent leaders known as "high value targets." The question is not whether small drones are useful, it's how useful they are compared to manned aircraft.

Money brings the issue into sharp focus. Small drones, notably the DJI Phantom series, are already transforming television and movie making by providing stable camera platforms at bargain basement rates. These are bigger versions of the Micro Drone, but it costs less to buy one outright than to hire a helicopter for an hour, and they can film in urban canyons and other places where no helicopter can fly.

This book focuses on what a swarm of thousands or tens of thousands of small drones can do for a fraction of the price of a single F-35. The rate at which the technology is developing, and the way evolution favors small drones with their short generation times, is putting manned jets on the losing side. The technology developed for smartphones puts the big guns on the side of the small drones.

Ten years ago, I wrote *Weapons Grade: the Links between Modern Warfare and Our High Tech World,* about the high tech in everyday life that originated with the military. From GPS and the Internet to digital cameras and jet airliners, all sorts of gadgets were originally developed for war. The building blocks of modern electronics, integrated circuits, and microprocessors came from the defense sector. It used to be something of a cliché to say that whether it was computers or aircraft, the Pentagon was always twenty years ahead of what was commercially available. You could see the future by looking at what the military had.

In the last decade, the situation has changed. It's not all about iPhones, although Apple's ground-breaking smartphone and record-breaking profits are obvious indicators of strength. Smartphone sales have accelerated from zero in 2006 to over a billion smartphones shipped in 2013. Calling these devices phones is deceptive: each has more computing power than previous desktop computers, as well as a digital video camera, GPS navigation, digital communications, and a stack of other sensors and gadgets built in.

The mobile phone industry has the power and momentum of a freight train. Billions of dollars are spent annually on advancing technology just for small electronic devices. Aggressive schedules see a new generation appear every two years. New types of software and hardware are constantly emerging. A phone that merely delivers incremental improvement over the previous generation is a letdown; every phone is expected to be astonishing.

In the course of this market-fueled acceleration, consumer electronics have outstripped their military counterparts. These days soldiers are less likely to be awestruck at the gadgetry they are issued than shocked by how clunky it is compared to the sleek lightweight devices they have at home.

Defense contractors argue that their products cannot be compared with consumer electronics. They have to comply with demanding standards that smartphones are not subject to – rugged enough to survive the battlefield, able to withstand high and low temperatures, must not create electronic noise and interfacing with existing military systems. Selling to the military means extensive testing and certification, with the related delays and costs. Add to this a military bureaucracy that can take years to agree on the specification it wants in the first place, overseen by a political leadership that may cancel, delay, or divert any project depending on the shifting sands of expediency, and you have a recipe for a long time between generations.

Each generation of electronics roughly translates to a doubling of processing power, memory, pixels, or other relevant metrics. If a commercial product goes through a generation every two years, and the military cycle takes six years per generation, then in twelve years the military product goes from being four times as powerful as the competition to a quarter as powerful. As a result, the prospect for the military afterwards is watching disappearing taillights as their rivals pull ever further ahead. The military has recently started taking the obvious course of adapting commercial electronics rather than developing their own.

Smartphone technology has made electronics smaller, cheaper, and more capable than ever. Technologies developed for phones fit well with the requirements for small drones. Like phones, drones need miniature cameras, GPS navigation, and data processing power. Both share the same need for minimal size, weight, and power. A drone is simply a smartphone with wings, and the wings are the cheap part.

The stage is set for small military drones made with off-the-shelf electronics. Drones that will be cheap and plentiful, in contrast to the current generation that cost tens of thousands of dollars apiece. And they are getting more powerful all the time. These are not just dumb, remote control aircraft, but smart drones with a degree of autonomy.

A swarm of armed drones is like a flying minefield. The individual elements may not be that dangerous, but they are so numerous that they are impossible to defeat. They can be disabled one by one, but the cumulative risk makes it safer to avoid them than to try to destroy them all. Minefields on land may be avoided; the flying minefield goes anywhere. When it strikes targets on the ground the swarm can overwhelm any existing opposition by sheer numbers of intelligently-targeted warheads.

Three key characteristics – its robustness, low cost, and rapid evolution – could make the drone swarm the ultimate weapon.

Three main objections are often raised regarding the idea that thousands of small drones could dominate the battlefield, and a chapter is devoted to each.

First is the issue of power: with battery life of only an hour or two, how can small drones stay in the air long enough to achieve anything strategic?

Second, the control issue: if one operator is normally required for every drone, how is it feasible to fly thousands of them?

Third, the issue of firepower: how can anything so small make a real impact?

The answer to all three questions lies in existing technology. Future drones will be able to draw energy from the environment and recharge themselves, and thousands of them will operate as a single swarm requiring just one operator – or none. And new developments in smart weaponry make swarms of small drones more lethal than anything currently available.

There are high entry barriers to the military aircraft market. When the F-35 was being developed, there were only two consortia big enough to compete for the contract to build it. Few can keep up in a business where it costs billions to bring a product to market in a process that may take decades. This does not apply to small drones. Not only can the smallest of companies join in, university departments and hobbyists are building their own drones, often with innovative features. Advanced design software and 3-D printing mean that for a few thousand dollars a new concept can go from fleeting thought to flying hardware in days. The F-35 is edging slowly towards service; meanwhile the Micro Drone 2.0 is already being replaced by the 3.0 version boasting advances such as streaming HD video.

Small drones represent an unparalleled opportunity. Military operations can be carried out with greater precision, with less collateral damage and less risk to friendly troops than ever before. Small drones present a genuine challenge to the idea that we will always need "boots on the ground" in a military conflict. And they can do the job cheaply. They are not just cheaper than planes, they cost less than the current missiles, bombs, and fuel expended in airstrikes.

Small drones also represent a serious threat to the West. For reasons we will explore, existing weapons cannot stop them, and the flying minefield is likely to dominate in the air and on the ground. In the hands of hostile forces, a drone swarm could inflict massive damage – not just on the battlefield but against our cities.

Small drone evolution means that the most powerful military in the world is no longer in control of the future of airpower, a future which will not belong to the F-35. Rather, the ultimate weapon could be in the hands of anyone with access to smartphone technology. This includes other countries, as well as non-state actors like ISIS and Hezbollah or even activists and hacker groups like Anonymous. The only way to stop an attack from a smarm may be with another swarm – if you have one.

This is the world of Swarm Troopers. How we have arrived here, where we are going next and what the possible consequences are will be laid out in the chapters that follow.

Structure of the Book

Swarm Troopers describes the history and background of drones before covering recent developments and then looking at where the technology is heading.

Chapter One: "Drone Prehistory: War against the Machines" is a brief history of the drone, highlighting some of the most important unmanned aircraft of the past and why they failed to gain any traction in spite of their successes. Plus, a small footnote in history, the Japanese "windship weapons," which struck at the US mainland from across the Pacific in WWII.

Chapter Two: "The Surprising Rise and Sudden Fall of the Predator" outlines the big breakthrough of the first truly successful drone. How the Predator emerged from the CIA's classified Amber program and evolved from being a highly effective flying spy to an international killing machine. We also look at why it is now under threat as the Pentagon leans away from unmanned programs.

Chapter Three: "The Raven – a Small Revolution" looks at the little drones that have been overshadowed by their larger cousins but are numerically far more prevalent. How the Pointer, a primitive and impractical drone, evolved into the Raven, an essential tool for the modern soldier that accounts for 90% of the Pentagon's unmanned fleet.

Chapter Four: "A Sharp Fall in the Price of Victory" discusses why military aircraft are so expensive and why small drones are different. Why drones are cheap and getting cheaper, and how a team from a US think tank delivered 90% of the capability of existing military drones at 10% of the cost. How 3-D printing and the smartphone revolution are being harnessed to produce innovative drones.

Chapter Five: "The Fine Art of Flying Forever" looks at how the limited flight times of small drones are being extended by energy harvesting technologies. How drones can run on solar power or land on power lines and recharge themselves from the grid – or tap the energy of the wind itself to fly indefinitely like soaring birds.

Chapter Six: "Faster Forward" looks at the trends in technological development, and how small drone evolution is accelerating. How the drive for better smartphones keeps spinning off new technologies that benefit drones, and how the rise of "biomimetic" technology produces drones that resemble living things, making them as capable as birds.

Chapter Seven: "Teams, Formations, and Flocks – the Power of Swarms" explores research into the science of swarming behavior, looking at how groups can operate together to become more than the sum of the individuals. How a few simple rules can meld hundreds or thousands of birds, insects, or robots into a single, coordinated unit capable of carrying out tasks so a single controller can handle thousands of drones. Why swarms are smarter, tougher, and deadlier than uncoordinated groups.

Chapter Eight: "Miniature Terminators– The Power of Small Weapons" looks at how much damage small drones can do. From anti-personnel weapons to miniature demolition devices and high-temperature incendiaries, new technology is making small drones into true engines of destruction.

Chapter Nine: "Fighting the Swarm" looks at ways of countering the threat of swarms of small drones, why existing weapons are ineffective, and what the alternatives are, including missiles, directed energy weapons, and EMP. How the real answer may lie in the Navy's "aerial combat swarm" contest, which seeks to counter to a swarm of drones with another swarm of drones.

Chapter Ten: "What Happens Next?" While the potential of the weaponized drone swarm may seem obvious, both as a threat and as a new direction of the military, history does not necessarily indicate that it will be adopted. This final chapter explores what the likely outcomes will be, who are likely to be the winners and losers, and how the new opportunities and threats posed by drones might be addressed.

CHAPTER ONE

DRONE PREHISTORY: WAR AGAINST THE MACHINES

> *"I could throw my bloody umbrella further than that!"*
> - Major Bell, British army officer at an early drone demonstration in 1916

The history of drones has not been one of steady progress. It is the story of a war, in which every advance made by the machines has been checked and driven back by opposing forces. Again and again, technology has emerged and unmanned aircraft have taken off briefly, only to be brought down again. Like vengeful Roman emperors, the victors have erased even the memory of those they have destroyed. History shows that drones tend to be ruthlessly terminated by a military establishment that harbors a ferocious antipathy to anything that dares to compete with manned aircraft.

The politics are simple. Manned programs have plenty of friends in high places; drones have none. The objections to drones are equally simple. Manned aircraft are proven, reliable, and well understood; drones are none of these things. Inevitably, the technology cannot match human performance in every way. In the cut-throat business of military procurement, such weaknesses are easily exploited.

The US Air Force in particular has been accused of "white scarf bias," meaning that senior officers tend to be ex-combat pilots. They see things from a pilot's point of view. They value piloting skills and have the greatest respect for fighter aces. The dream of flight is the dream that has carried them along since childhood. They have little enthusiasm for "pilots" who sit on the ground at a computer console and unmanned aircraft that are inferior to manned planes. Why would they support drones?

Seen with the benefit of hindsight, any opposition to technology can look foolish. This may be deceptive. The cavalry officers who could not see the advantages of switching to motor vehicles at the start of the twentieth century were making a rational assessment based on the available evidence. Horses had centuries of successful service, while their motorized replacements had always been clumsy and unreliable, especially on rough going. There were various failed attempts to build combat vehicles before tanks arrived in 1916; Leonardo da Vinci may have had the idea first, but nobody could make it work. Until a vehicle could cross miles of muddy field without breaking down or getting stuck, "land ironclads" were a science-fiction fantasy.

It is not a simple objection to technology, as witnessed by the way the military took to the first aircraft. In 1907, just four years after their first flight, the Wright brothers were building an aircraft for the Army Signal Corps, when some still considered claims of powered flight to be unproven. By 1911, Italian aviators were dropping bombs in Libya, and aircraft were used by every major power in WWI. Aircraft have continued in military service around the world ever since. The military has embraced advanced concepts: helicopters, vertical takeoff tilt-rotors, and rocket planes, frequently when the technology was immature or downright dangerous.

Unmanned aircraft have a history almost as long as their manned counterparts. But it is not a continuous history. After each rediscovery, in the First and Second World War, in Korea, in Vietnam and beyond, the drones were knocked back into obscurity. Even when they have proven successful, nobody wants them when the war is over: "The great broom of victory swept all new projects into the ashcan of forgotten dreams," (2) as Commander Delmar Farhney put it. Farhney, whose little-known drones saw combat in WWII, had more cause to be bitter than most, as we shall see.

Drone prehistory goes back to 1849 and the use of bomb-dropping balloons in the siege of Venice. (3) These were devised by the ingenious Lieutenant Uchatius of the Austrian army. It was not possible to bring siege artillery close enough to the city. Uchatius, better known to history as a photographic pioneer, rigged up hot-air balloons to release small bombs by remote control via a copper wire. About twenty of the balloons were launched. Austrian news reports suggested the bombs would turn Venice into rubble, but it seems that only one or two hit the city. The rest fell into the waters of the Venetian Lido or outside the city entirely. There was no follow-up on the balloon-bomb project –at least, not for several decades.

In some ways the true inventor of unmanned warfare was Nikola Tesla, who demonstrated a miniature boat controlled by radio waves at Madison Square Garden in 1898. Tesla believed that a version armed with torpedoes could sink battleships and lead to a new age in which wars were fought between machines with no human combatants. As with many of his projects, Tesla never developed the idea beyond the initial demonstration.

Both Britain and the US developed their own drone aircraft in WWI. The British effort was headed by "Professor" Archibald Low – he used the title even though he was not actually a university professor. Low had a tremendous enthusiasm for remote control and was repeatedly distracted from one project by another. His project was known as "AT," short for Aerial Target, a designation to mislead the enemy into thinking the device was simply a target for anti-aircraft practice. This name proved to be strangely prophetic.

The AT was a wooden biplane with a fourteen-foot wingspan and an explosive warhead, intended for use against both ground targets and zeppelins. It flew well initially, but the program was terminated after an unfortunate incident in 1917 when it was being demonstrated to a group of generals. Low enjoyed being in the limelight and was good at attracting publicity. When it came to the AT demonstration, though, Low's enthusiasm for the fledgling craft led him to disastrously overestimate how ready it was for public display.

The first AT to be launched in the demonstration suffered engine failure during take-off and flopped into the ground. A Major Bell delivered the verdict quoted at the head of this chapter: "I could throw my bloody umbrella further than that!" (1)

The second machine fared even worse. The operator lost control and the AT flew right at the audience, scattering them before veering off and crashing a few feet away. As a demonstration of controlled flight, it was unconvincing. Nobody present can have thought that it was advisable to put high explosives on drones.

The American effort was headed by Elmer Sperry. Sperry is best known as the inventor of the gyrocompass, a non-magnetic, spinning compass that gets its direction from the rotation of the Earth. Sperry also invented a gyrostabilizer working on similar principles, which would keep an aircraft stable without human control. By 1918, Sperry's Aerial Torpedo was able to fly along a preset route and dive on a target, delivering a thousand-pound bomb or releasing a torpedo. The weapon was too late to be used in action, and after the war, the US Navy thought drones were useful only for gunnery practice. The US Army developed a similar aircraft, called the Kettering bug, but showed a similar lack of enthusiasm.

Even when relegated to the ignominious role of targets, drones developed a knack for embarrassing humans. In the 1930s, the British Air Ministry decided to test the claim that battleships were vulnerable to air attack by flying a radio-controlled Fairey Queen target plane against the British Mediterranean fleet. After more than two hours of sustained anti-aircraft fire and numerous passes, the drone was undamaged. (4) The Royal Navy accepted that its air defenses needed upgrading. Large numbers of drones were built as a result – but only as targets. Nobody thought the test showed that an unmanned aircraft might be an effective weapon.

World War II: America's First Attack Drones (5)

Lt. Commander Delmar Farhney worked with the US Naval Research Laboratory in the 1930s building radio-controlled anti-aircraft targets for the Navy. It was an exciting era to be working with radio, and Farhney was convinced that unmanned aircraft would be devastatingly effective. By 1941, he had extended his work to aircraft capable of accurately dropping torpedoes and depth charges. Incidentally, Farhney was the first to officially refer to his aircraft as "drones," a usage the military has since tried to suppress.

At the outbreak of war, Farhney was ordered to develop and produce an assault drone as soon as possible. Previously, obsolete manned aircraft had been converted into drones, but this was not an option for Farhney, as every available plane was being used for training or combat. In addition, Farhney was told that his work should not use any resources needed for manned aircraft production. Rarely has the conflict between manned and unmanned projects – and the relative status of the two – been so clear.

Farhney built a new type of aircraft from scratch, one that would be small and fast and make a difficult target. He could not use metal, so the TDN-1 was made from plywood. Successful flight trials from the aircraft carrier USS Sable in 1943 proved the principle, but design changes were needed. The follow-on was the TDR-1, made by Interstate Aircraft & Engineering, another masterpiece of plywood and improvisation. (The "TD" stood for "Torpedo Drone", with a suffix to indicate the maker). Some of the work was carried out by organ makers Wurlitzer, with their long experience at shaping plywood. The TDR-1 had a wingspan of forty-eight feet, a speed of almost a hundred and fifty miles per hour, and awkward tricycle landing gear to give space for a 2,000-pound bomb or torpedo slung beneath the fuselage

As an airplane, the TDR-1 was unremarkable, but it was equipped with a remarkable technological breakthrough: remote control by television. Dr. Vladimir Zworykin of RCA was one of the inventors of the television, and he was keen to put it to use in drones. The prototype cameras weighed over three hundred pounds including the transmitter, but this was shrunk into a miniature system weighing ninety-seven pounds, packed into a box the size of a carry-on suitcase. The picture was monochrome with a respectable resolution (350 lines) and a refresh rate of forty hertz, but the image was poor by modern standards. The drone operator had to work under a black cloth to see the green, five-inch screen clearly in daylight.

The TDR-1 also had a novel radar altimeter to maintain flight at a steady altitude. This helped make TDR-1 stable and easy to fly by remote control. The plane had its own quirks, though, including a lack of brakes on the landing gear. It was controlled from a modified Avenger torpedo bomber flying up to eight miles away. The special Avenger had a crew of four, with pilot, radio operator, and gunner joined by a drone operator. The latter had a joystick, a television screen, and a rotary telephone dial. The dial controlled altitude and released weapons by dialing specific numbers, and the television gave a real sense of being in the drone.

"To sit under the hood of an airplane and control the one up ahead with radar and television, where you can actually see where you are going – it felt like you are actually flying that plane," said one operator.

The drones were formed into a unit, Special Task Air Group One or STAG-1, but the Navy was reluctant to deploy this untried weapon. The drones were eventually allowed to attack a derelict Japanese freighter called Yamazuki Maru off Guadalcanal. Three out of four drones hit the target, and, after some hesitation, the unit was sent into action.

The STAG-1 drones successfully attacked anti-aircraft sites, gun positions, ships, and even a lighthouse. Many of them were used in suicide attacks against challenging targets; the Japanese, not knowing they were unmanned, called them "American kamikazes." Other TDR-1s were armed with a mix of smaller bombs, sent to attack multiple targets and return.

"I distinctly remember the excitement, watching the grainy and sometimes static-filled green TV screen as the drone I was guiding approached the grounded ship. When an unfamiliar pattern of small dots began to appear, I thought the receiver was malfunctioning. Suddenly I realized; they were flak bursts! But I kept the drone on target, concentrating on holding its bouncing nose squarely on target. I crabbed it a bit to correct for wind and to avoid the worst of the flak. At the last second, I had a close-up view of the ship's deck. Then…just static. I had hit the ship squarely amidships."

Though hard to hit, some of the drones were brought down by anti-aircraft fire against heavily defended targets. "Yeah, I got shot down once or twice," recalled one operator.

The commander of the STAG-1, Lt Commander Robert Jones, was convinced that their successes would prove the value of the drone concept. He believed drones would be an important weapon in the assault on mainland Japan. But the Navy top brass did not agree. The drones might be good for precision attacks, but what were needed were formations of heavy bombers. After the drones were all expended, STAG-1 was reassigned. Commander Jones watched unhappily as the thirty Avenger control planes were dumped overboard in Reynard Sound.

The television technology was a limitation, as anyone who has worked with monochrome images can appreciate. Targets that had a clear silhouette, like a ship on the water, showed up clearly and were easy to hit. But any target surrounded by jungle tended to be invisible on the small screen, as they blended in with the confused background.

(Meanwhile, experiments with larger radio-controlled aircraft as suicide bombers against major targets had limited success. In the most famous disaster, Lt Joseph Kennedy Junior was killed when a "robot" PB4Y-1 bomber blew up prematurely in 1944. This left his younger brother John F Kennedy as the family heir.)

Both Farhney and Jones continued the struggle to get drones recognised, and during the Korean War, unmanned attack planes were tried again. In 1952, six obsolete F6F Hellcats were converted to unmanned operation. They were controlled from nearby AD-2Q Skyraiders, with a television system developed from the one on the TDR-1. Flying from the aircraft carrier USS Boxer, the drones successfully hit a power plant, a railway tunnel, and a bridge. Jones wanted to continue operations and attack the Yalu River bridges, which had survived repeated attacks by US heavy bombers.

This might have been a moment for technology to prove itself, like the mission to destroy the Thanh Hoa Bridge in Vietnam exactly twenty years later, which we will explore in Chapter Eight. But the Navy brass did not approve the plan and Jones never had his chance with the Yalu River bridges. The idea of attack drones was again shelved, and unmanned aircraft were once more relegated to the role of targets.

Farhney went on to become a Rear Admiral and headed the Navy's guided missile research effort, but he was never entirely at one with the establishment. In the 1950s he made a number of public statements about UFOs, which he believed to be craft of extraterrestrial origin. At least his drone experience had given him plenty of practice at dealing with unbelievers.

DASH: The Little Helicopter That Could (6)

The QH-50 DASH was an amazing little helicopter decades ahead of its time. Like many drones, it was a good enough machine but fell foul of politics.

In the late 1950s, the latest sonar could detect a submarine more than twenty miles away, but the best anti-submarine weapons only had a range of a few miles. The US Navy wanted to bridge the gap with a Drone Anti-Submarine Helicopter or DASH. This was a small helicopter capable of carrying a single weapon and dropping it at the required spot, guided by a controller back on board ship.

The DASH was based on a one-man helicopter called a "Rotorcycle" built by Gyrodyne Company. This had two rotor blades rotating in opposite directions for lift, and a propeller for forward motion. The drone version was the size of a small car and weighed just over a ton. By 1963, the US Navy had eighty of them.

There was political opposition though. Politicians know that attacking expensive and wasteful government spending is always a popular move, and some, including Congressman Yates from Illinois, zeroed in on what appeared to be unfavourable reports on the DASH.

By 1967, Secretary of Defense McNamara was expressing doubts over the program because, he claimed, DASH had a higher-than-expected loss rate. This was not exactly true. There were problems with training and a lack of planning, which caused some losses, but the real problems were organizational. DASH was designed to be expendable; when it dropped a Mk57 nuclear depth charge it would be within the lethal radius of the resulting explosion. The powerful warhead, from five to twenty kilotons, guaranteed that the sub would be destroyed, and losing one drone for one submarine was a good exchange rate. The idea that DASH should carry a non-nuclear homing torpedo and come back afterwards was a case of mission creep; according to the original design it was only supposed to make one flight.

The average flying time before a crash was originally set at eight hours, but by 1968 the loss rate had been reduced to one loss per sixty hours flying. The later model DASH was even more reliable. The situation improved when maintenance crews identified a condition known as "purple plague," a form of corrosion affecting cheap transistors. They cured it by substituting better quality components.

Although made with off-the-shelf components rather than specially tailored ones normally used in military aviation and not meant to last, DASH survived repeated use. But it was under threat.

The Vietnam War was expensive, and anti-submarine weapons were not a pressing requirement.

This spurred the Navy to find new uses for their drone. Executive Officer Phil King of the USS Blue modified a DASH, adding a television camera for reconnaissance and gunnery direction. Known as SNOOPY missions, these involved the DASH flying out to find targets. The operator identified them via the television link, and the destroyer then opened up with its battery of five-inch guns. The drone operator could see where the shells were landing and tell the gunners how to adjust their aim.

Further developments followed, including NITE PANTHER and BLOW LOW versions equipped with additional fuel tanks for longer range, night-vision systems and airborne radar.

The next logical step was to convert the DASH from finding targets to attacking them. NITE GAZELLE, GUN SHIP, and ATTACK DRONE were all individual modified aircraft with a range of weaponry including a six-barreled minigun firing four thousand rounds a minute, grenade launchers, bomblet dispensers and bombs, as well as a laser designator for directing smart bombs. The idea was that drones with guns would deal with the ground defenses, leaving the way clear for the bomber drones to hit targets with pinpoint accuracy.

Some aspects of the DASH projects are still classified. Known modifications included a surveillance version, a machine for rescuing downed pilots, and even one for laying smoke screens. There may have been others specially modified for particular missions. It was undoubtedly a highly versatile little aircraft. Some reports even claim NITE GAZELLE carried out night attacks on North Vietnamese convoys in field testing and evaluation, but this remains unproven.

The political assault on the DASH continued, including more allegations from Congressman Sidney Yates that the loss rate was too high. The media repeated Yates' claims, and the Navy decided not to defend DASH. Internal documents released later show that the Navy commanders still believed in DASH but were more concerned about a bigger project. This was reflected in a comment from Captain Chris Zirps, who noted that in spite of the steadily improving reliability of DASH, he was still getting negative feedback from his superiors: "It became quite evident that the Navy no longer wanted DASH and wanted to move onto LAMPS manned helicopters."

LAMPS was the Light Airborne Multipurpose System, a new manned helicopter that would operate from destroyers and take over the role of DASH. Removing DASH from the picture meant there would be no competition, and nobody would be able to argue that LAMPS was unnecessary.

DASH's supporters – chiefly those who actually worked with the little drones – still maintain that DASH was betrayed (7). The LAMPS project became the SH-60 Sea Hawk, now a multibillion dollar success story. Although, as one DASH operator notes, "When DASH was shot down on a mission or crashed at sea, you ordered another drone from a Destroyer Tender. When a Sea Hawk crashes today, you have six funerals to arrange."

The DASH lobby also notes that US Navy ships still do not have any unmanned capability, making them in at least one way less able than their 1960s counterparts.

When DASH was cancelled in 1970, there was the usual question of what to do with the remaining aircraft. Being drones, they were consigned to the usual fate: they were sent to the White Sands Missile Range as targets. When numbers started to get low, the surviving DASH were used to tow targets instead, still a fairly hazardous occupation. The last one was shot down some time in the 1990s and evidently little mourned.

The Fighting Fire Flies (8)

In contrast to the DASH, which started out as a combat aircraft and ended as a target, the Teledyne Ryan Firebee was born to be a target, but showed it was capable of much more. In the process, it attracted hostility from those who prefer their planes to have pilots in them.

In 1960, the risks of using manned reconnaissance planes over enemy territory were all too clear. An unmanned aircraft could perform the same mission, taking pictures of a target area and returning with no risk of a pilot being shot down and captured. A suitable airframe already existing in the form of the Firebee series of target drones made by the Teledyne Ryan company.

The Firebee was fly at any height from the treetops to fifty thousand feet. It could be launched from an aircraft and remotely controlled from two hundred miles away. The Firebee would return to the ground on a parachute, an easy feat for a small plane with no human inside risking broken bones.

There was little interest from the Air Force's mainstream, but the highly unconventional BIG SAFARI team liked the idea. BIG SAFARI was set up to circumvent the usual complexities of Air Force procurement, to provide quick solutions to urgent problems. They funded development of a version of the Firebee called Fire Fly or Model 147, and it went through their streamlined channels without the interference it might have otherwise endured.

There was considerable reluctance to deploying the Fire Fly, and the Air Force decided not to use it over Cuba, preferring to risk manned U-2 spy planes. During the Cuban missile crisis in 1962, Rudolf Anderson was flying a U-2 over Banes in Cuba when a Russian-made surface-to-air missile scored a near miss. Shrapnel punctured Anderson's pressure suit and he died of decompression, making him the only combat casualty of the Cuban crisis.

In spite of this and other losses, nobody wanted to try the Fire Fly. After one presentation, a senior officer told Fire Fly advocate Colonel Fred Yochim flatly, "It will never replace manned reconnaissance."

The Vietnam War should have been an ideal opportunity to try out the new technology. However, there was the old problem of persuading commanders to give Fire Fly a chance. Operational tests were staged to establish whether the drone would be effective against air defenses. Unfortunately, this meant flying against Air Defense Command off Florida. Either the drones would be shot down and fail, or they would get through and the defenders would fail. Somebody had to lose.

In the first trials the F-102 Delta Dagger and F-106 Delta Dart pilots never even saw the drones they were trying to shoot down, and only caught brief glimpses of them on radar. Further tests followed. In one, a Delta Dagger fired a burst of cannon fire at the drone, but the rounds missed. Before the pilot could line up for another shot, his jet engine flamed out because of the high altitude. He dropped to lower altitude to reignite the engine, at which point other planes mistook his aircraft for the target. Fortunately, they did not shoot, but the Fire Fly had escaped. Later on two Delta Darts achieved a radar lock on the Fire Fly, but not for long enough to fire a missile.

Honor was at stake. Air Defense Command assigned their best pilots for the next tests, and they finally shot down some drones. In one test, a US Navy pilot scored a near miss with a missile, triggering the Fire Fly's recovery parachute and sending it spiraling out of the sky. The pilot returned to make another pass and launched a second missile to destroy the stricken drone, apparently out of sheer spite. Shooting at human pilots parachuting down is a violation of the Laws of War, but clearly, machines do not count.

The military was unhappy with the results. Many felt the test was intended to make them look bad. Robert Schwanhausser of Teledyne Ryan says the results were classified Top Secret, and he was ordered to burn every piece of information on them. But the drones had made their point.

The Fire Fly drones went on to perform well in Vietnam, repeatedly photographing targets that were considered too dangerous for manned aircraft. They were sent on virtual suicide missions, to test Vietnamese radar and missile defenses.

When losses mounted, the developers at BIG SAFARI started equipping their drones with electronic bags of tricks. One device, known as High Altitude Threat Reaction and Countermeasure (HAT-RAC) responded to being lit up by radar by throwing the drone into a series of sharp turns. Others carried out preprogrammed evasive maneuvers to evade possible surface-to-air missile attack.

The drones were flown on secret spy missions over China, an enemy of the US in the Cold War. They proved very difficult to stop. When the Chinese downed their first Fire Fly in 1964, it was only after some sixteen MiGs had made over thirty passes trying to hit the little drone.

Fire Flies were even credited with bringing down a number of Vietnamese MiGs. In some cases, a pursuing jet ran out of gas – high-speed chases can drain an aircraft's tanks in minutes. In others, a MiG was confused with the drone and fell victim to friendly fire.

A decoy version of the Fire Fly was produced. This was known as the 147N and was fitted with radar reflectors to make it look like a bigger aircraft. The 147Ns were originally purely intended to distract defenders away from the real Fire Flies equipped with cameras, but they survived and managed to return so frequently that they were later fitted with cameras of their own.

As well as carrying out high-altitude missions, some Fire Fly drones were modified for low-level photography flights. Sometimes extremely low: in one case, the plane apparently flew underneath high-tension electric cables.

"We ended up getting stuff that was just unbelievable," recalls a flight engineer. "I got into the photo recon center in Saigon and saw some of the material that was coming back. It was really amazing looking at the detail of shops there in Haiphong Harbor."

The drones were used for high-priority, high-risk missions such as finding out where US prisoners of war were being held.

On one mission, the pictures from a Fire Fly captured the subject's faces from close range: "You could see features on the guy's face. If it would have been in color, you could have seen the color of his eyes."

This was at a time when the U-2 spy planes were taking pictures from fifty thousand feet or higher, with resolution only good enough to recognize objects two feet across. The low-level Fire Fly pictures were a revelation in the art of the possible.

Another version of the Firebee showed how formidable drones could be in a dogfight. The basic drone could only handle acceleration of about 3G, but a modified Firebee equipped with "Maneuverability Augmentation System for Tactical Air Combat Simulation" or MASTACS could pull 6G for several seconds at a time. This put it pretty much on a par with manned fighters. In 1971, the MASTACS developers challenged Commander John C. Smith, head of the Navy's Top Gun combat training school – the "Top Gun" of the 1982 movie – to try and shoot MASTACS down.

Smith and his wingman, both flying F-4 Phantoms, made repeated attacks on the remotely controlled Firebee. It was far too agile for them. They fired two Sparrow radar-guided missiles and two Sidewinder heat-seekers without scoring a hit. Meanwhile, the Firebee kept circling around and lining itself up in firing position behind the Phantoms. Had it been armed, the Firebee would have had easy shots.

As usual, nobody liked a smart robot. MASTACS was deemed "too sophisticated" for training purposes. The Navy did not have any requirement for an unmanned fighter, and another robotic upstart went down in metaphorical flames.

It is not surprising to learn that the proposed *Top Gun* movie sequel will see Tom Cruise and friends addressing the question of whether human pilots are obsolete in the age of drones. Anyone expecting a win for the unmanned aircraft may be disappointed.

Other proposed versions of the Firebee suffered similar fates, in spite of successful demonstrations. A ground-attack Firebee called Model 234 was developed to take out surface-to-air missile sites after the missiles proved deadly to American aircraft in Vietnam. In a test at Edwards Air Force Base in 1971, the 234 launched a guided missile at a van representing a radar control vehicle and scored a direct hit. Further successful tests with missiles and guided bombs followed, but the Air Force decided there was no requirement for killer drones. Attacks on missile sites would be carried out by manned planes – a task so dangerous that one critic likened it to "charging at an elephant gun with an elephant" because it deliberately exposed the aircraft to something specifically designed to shoot it down.

The end of the Vietnam War brought US drone development to a halt. In spite of multiple successes, the cause of unmanned aviation had not advanced. The reconnaissance Fire Flies were retired, and other projects were dead ends. The only Firebees that remained were targets for pilots to hone their skills on – although some were also exported to Israel, where they found other uses as we shall see shortly.

Even the memory of the Fire Fly seems to have been lost. In 2014 the US Navy proudly announced in a press release that, "Truman will be the first aircraft carrier in naval aviation history to host test operations for an unmanned aircraft." It seems that amnesia buried the 1969-70 Fire Fly operations from the USS Ranger, not to mention the TDR-1s flown from the USS Sable in 1943.

The Aquila Fiasco (9)

The foregoing gives an indication of the Air Force and Navy's distaste for drones. It is understandable why they might see them as a threat to pilots and react accordingly. More surprisingly, the Army, a service with less stake in manned aviation, has also been slow in accepting drones. There have been numerous unsuccessful developments over the years, but the most notable gravestone by the wayside is surely that of the MQM-105 Aquila.

The Aquila was a tactical drone developed for the US Army in the 1970s and early 1980s. For once, it was a victim of management failure rather than political opposition.

The Israelis had shown that small, cheap drones could be tactically useful. In 1982 Israeli drones fitted with TV cameras located Syrian surface-to-air missile radar, while other drones carried radar jammers or acted as decoys. A squadron of Firebees mimicking fighter jets tempted the surface-to-air missile units to turn on their radar and reveal their location; the Firebees evaded every single one of the forty-three missiles fired at them. The defenders were left vulnerable to a follow-up strike by manned aircraft before they could reload. Using this combination of drone tactics, the Israelis destroyed seventeen missile sites with no loss. Drones, it seemed, might really be useful if used properly.

Aquila was partly a response to the Israeli success. Aquila would give a soldier a view of the other side of the hill, and would be able to direct artillery fire without the need for an observer on the spot. It also provided a new, high-tech means of tackling the Soviet tank divisions massed on the border between East and West Germany. Artillery was vastly more effective against armored vehicles thanks to new "bomblet rounds" that scattered the area with hundreds of armor-piercing mini-bombs instead of a single warhead. However, an observer still had to make sure that shells were landing in the right area, calling corrections if the aim point needed to be shifted.

There was also a brand new laser-guided artillery shell called the M712 Copperhead, which could knock out a tank from ten miles away with the first shot – but there had to be an observer on the scene with a laser designator to illuminate the tank.

The Army's answer was the Aquila, a drone with a stabilized TV camera and a laser designator to mark targets. It looked like a miniature stealth bomber, with a twelve-foot wingspan and a wooden propeller driven by a modified go-kart engine. Aquila was launched by hydraulic catapult from the back of a special truck, which was also equipped with a net to catch the drone when it returned.

The project was not managed well. Aquila went from being a cheap and simple drone to a "gold-plated" one with every modern development. The Israeli drones cost around $40 thousand each; Aquila started out at $100 thousand and went up rapidly from there.

Specifications were changed repeatedly and grew increasingly demanding, and the project became a spectacular demonstration of the dangers of feature creep. Instead of following the initial plan of fielding a simple system and adding upgrades later where possible, everything had to be included from the start.

Aquila needed to be stealthy, which demanded an elaborately shaped body, limiting space inside. The cheap daylight TV camera was supplemented with an expensive thermal imaging camera. Communications were made jam-proof with the aid of complex steerable antennas and state-of-the-art radios that fired off data in short bursts. It gained a sophisticated navigation system: in the days before GPS, this was an inertial measurement system based on gyroscopes, a sort normally fitted to manned aircraft. It all pushed up the cost.

Costs attracted more costs. In order to ensure that expensive drones were not lost, Aquila had an automated recovery system using infra-red sensors and beacons, supplemented with an emergency parachute.

On top of this, the whole thing was hardened to withstand the effects of a nuclear blast. Aquila really was designed for World War Three. By 1984 the sticker price was somewhere over a million dollars per aircraft.

Still the military kept asking for more, and the contractors were happy to keep supplying it and adding to the bill. By 1988 program has absorbed around seven hundred million dollars with only a few dozen Aquilas built, and with technical issues still to be resolved.

The project might yet have been saved. Cost overruns and teething troubles are routine features of military procurement. With good management and a more realistic approach to the specifications, Aquila might have been saved. But the political situation was against it. The drone was never popular with Army aviation, and it was in direct competition with the Advanced Helicopter Improvement Program (AHIP).

The OH-58 Kiowa scout helicopter lacked several of the features needed by the Army – night vision, a stabilized camera, and a laser designator, things the Aquila had. Under AHIP, the OH-58 would become the OH-58D with the addition of a ball turret mounted on a mast above the rotor blades. The turret carried stabilized optics with a laser designator, and its position meant that the Kiowa could see targets and mark them with its laser while the helicopter hovered low behind cover. In effect, the new AHIP would replicate many of the capabilities of the Aquila.

Congress needed to make budget cuts. Upgrading to the OH-58D was not cheap – at almost six million dollars apiece, it was far more expensive than the basic Kiowa. But it was popular with the Army's aviation command. The small drone with no supporters was squarely in the firing line. Aquila was duly terminated, and the Army was left with a bad experience of drones.

Nobody could understand why it was so difficult and complicated simply to put a TV camera on a remote-controlled plane. The failure of Aquila was a strong argument against further drone development for many years: "We tried them before, and they didn't work."

Spies Heart Drones

A handful of drones did eventually make it into service. Interestingly, this was also inspired by Israeli experience. Some US drones were acquired directly from Israel. These did not compete directly with high-performance manned planes, but were short-range, propeller driven craft with niche roles. Specifically, they were used by the Navy for spotting targets at long range – a role which the QH-50 DASH had carried out thirty years earlier.

The Hunter and Pioneer were both propeller-driven drones with a wingspan of less than twenty feet that resembled light aircraft. The Pioneer was a slightly modified version of an existing Israeli drone, the Mazlat Scout. The Navy used forty of them in the 1991 Gulf War, with twelve being lost in the process due to a mixture of accidents and enemy fire. The most memorable drone incident in that conflict was when some Iraqis attempted to surrender to a Pioneer; the Smithsonian Museum asked for the drone involved in this incident for their display. The Pioneer supported Marine operations for some years afterwards.

The Army's Hunter drone was less successful. In some distinct echoes of the Aquila program, this turned out to have inadequate range and an ineffective data link. Its size increased so much from the original specification that Hunter could not fit into a transport aircraft as planned. The projected cost for a planned fleet of fifty grew to $2 billion; each small, propeller-driven light aircraft cost a phenomenal $40 million.

The Hunter flew in Iraq and in former Yugoslavia. It was used for intelligence gathering in former Yugoslavia, with a feed from its video camera relayed back to NATO headquarters, paving the way for a more illustrious successor, but the Hunter never seems to have surmounted its initial shortcomings. It has been described as another victim of "specsmanship" – based on unrealistic specifications, it ended up trying to achieve too much.

If the Pentagon hates drones, the CIA seems to love them. Drones have a unique capability to carry out deniable operations, which are important to the CIA. The Agency learned the hard way just how disastrous it can be when a spy plane mission goes wrong.

Two years before the Cuban Missile Crisis, one of the CIA's U-2 high-altitude spy planes was shot down over Russia. The U-2 was designed to fly far above the height of a country's defenses, but the CIA knew it was only a matter of time before the Russians found a way of getting to it. The Grand Slam flights in 1960 were high-risk but strategically important.

President Eisenhower was told that the plane had been completely destroyed and the pilot was dead. A cover story was released that a NASA high-altitude weather research plane had accidentally strayed over Russia. The Russian premier Khrushchev accused the Americans of lying, announcing that they not only had the remains of the aircraft but "we also have the pilot, who is quite alive."

U-2 pilot Francis Gary Powers was paraded for the media in Moscow and apparently admitted to being a spy. At a summit in Paris two days later, President Eisenhower was humiliated when Khrushchev refused to open the talks until the US apologized for the U-2 overflights and punished those responsible. Eisenhower responded by revealing the existence of the U-2 program.

Drones pose no such risks and are conveniently deniable and expendable. Four years after the U-2 incident, the Chinese shot down a number of Fire Fly drones in their airspace. The remains were put on display and, like the Russians before them, the Chinese denounced American imperialist aggression. But there was no media interest. The Chinese might well claim that the peculiar wreckage was from American unmanned spy planes, but where was the proof?

There was none of the international outcry that had accompanied the Gary Powers incident and no embarrassment for the politicians or the CIA. Equally, there was no risk that the pilot would be interrogated and give away information. (The main long-term consequence was that the Chinese reverse-engineered the drones. They ended up with a clone called WuZhen, which kick-started their own unmanned aircraft effort).

The US intelligence community, including the CIA, has continued to use drones for covert missions. These have always been low-profile affairs, with details only being released decades later if at all. The continued appearance of downed drones over Iran, Pakistan, and elsewhere, incidents seldom officially acknowledged by the US, testifies to their continuing usefulness.

When drones did eventually find a place in the US military, thanks to the success of the Predator, it was only with considerable assistance from the CIA. We will explore this remarkable success story in the next chapter.

But before we leave the history books, we should look at another little-known project from WWII, and the occasion when the US became the first country to be targeted by unmanned intercontinental weapons.

Fu-Go "Windship Weapons" (10)

In December 1944, US military observers on the West Coast reported a wave of unidentified flying objects. On investigation, these were found to be paper balloons thirty feet across. They moved fast even though they were drifting with the wind. The balloons were filled with hydrogen and had a complex mechanical gondola. At first, they were thought to be weather balloons, but after reports of unexplained explosions, one was captured intact and found to be carrying incendiary bombs. This was the Japanese Fu-Go or "windship weapon."

Nobody could figure out where the balloons were coming from. It was assumed they were being launched from submarines a few miles off the coast or perhaps by spies operating within the US or even agents within internment camps. It was months before intelligence revealed they had flown all the way from Japan. The Japanese were taking advantage of a newly discovered natural phenomenon, the jet stream, a narrow ribbon of fast-moving air at high altitudes.

The Fu-Go has a right to be known as a drone rather than a simple unguided missile because of its ingenious method of altitude control. A clockwork mechanism controlled the release of a set of small sandbags around the rim of the gondola. Whenever the balloon fell too low, it dropped another sandbag. If it rose too high, which might cause it to burst, a valve vented a small amount of hydrogen. This control system meant it maintained height and stayed in the jet stream for the three-day journey across the Pacific.

Once it had expended the sandbags, the Fu-Go started dropping its load of incendiary bombs, one by one. The aim was to start forest fires in the heavily wooded regions of the Pacific Northwest. This would spread panic and divert resources from the war effort. The target was big enough that even this rough method of aiming had a chance of success.

The simple balloons were made from mulberry paper and vegetable glue. Nine thousand were assembled by Japanese schoolgirls, many working in sports halls normally used for sumo bouts. Some of the workers were so hungry that they ate the glue they were working with.

US analysts estimated the Fu-Go cost $200 each, at a time when a P-51 Mustang was $50,000. The little balloons were hard to intercept. There was not enough metal on them to show up on radar, and they were surprisingly fast at high altitude, making them difficult to catch. Only around twenty were shot down.

At least four hundred Fu-Go made it to America, scattered from Mexico to Canada. The number would have been greater but for a problem with antifreeze in the altitude control system. This was too weak and the altitude controls were apt to freeze up, leaving Fu-Go to slowly descend into the waters of the Pacific. The Japanese were not aware of this and did not correct the fault.

The Fu-Gos had little concrete effect, largely because they were launched in winter when it was too wet for serious forest fires. They might have caused far more damage in the dry summer months. There were some casualties though. A child on a picnic in Oregon found a strange object in the woods and picked it up; it exploded, and there is a memorial to the five children and one woman who were the balloon bomb's only known victims. The nearest they came to hitting a military target was when a Fu-Go damaged a power line and briefly cut off electricity to the Hanford plant manufacturing plutonium for the Manhattan project.

However, the US had to devote vast resources to countering the balloon threat. Squadrons of patrolling fighters were assigned to intercept the Fu-Go with, as has been noted, little effect. A special paratrooper unit, the 555th "Triple Nickel" Parachute Infantry Battalion were trained as "smokejumpers," parachuting firefighters. The unit was sent west and based in Oregon and California to handle forest fires. The 555th parachuted in to fight fires on many occasions – over twelve hundred individual jumps in all – though these were not necessarily caused by Fu-Go. In any case, taking the whole battalion, plus their transport aircraft and other equipment, out of the theatre of operations may have been enough to justify the whole Fu-Go effort on its own.

The Fu-Go demonstrated how numbers of small, unexpected adversaries can slip past conventional defenses. The British had a similar project, Operation Outward, which also had modest success, but was cancelled because of the risk to Allied aircraft. After the war, the US looked at balloons as a means of delivery of weapons of mass destruction. The E77 balloon bomb was similar to the Fu-Go, but delivered an anti-crop agent in the form of feathers dipped in a bacterial or fungal culture. Like the Fu-Go it was an imprecise way of hitting a large target, but 1954 tests suggested that balloon bombs would be effective. However, more orthodox approaches prevailed – chemical or bacteriological bombs dropped from heavy bombers.

The US also tested long-distance balloons for photographing enemy territory, but again balloons were edged out by manned aircraft. As always, the US military took more interest in high-performance manned aircraft than small, unmanned alternatives. The lesson of the Fu-Go was forgotten. It may be relearned the hard way in the years to come.

REFERENCES:

1) "I could throw my umbrella –"

Taylor, J. (1977). *Jane's Pocket book of RPVs: Robot aircraft today*. London: Macdonald and Jane's

2) "The great broom of history…"

Spark, N.T. (2011, June). Command Break by Nick Spark. (http://bit.ly/1HGU0oY)

3) "Drone prehistory goes back to 1849 and the use of bomb-dropping balloons in the siege of Venice."

4) "Remote Piloted Aerial Vehicles: An Anthology." (http://bit.ly/1NIdwEf)

5) "After more than two hours of sustained anti-aircraft fire and numerous passes, the drone was undamaged."

Newcombe, L. (2004). *Unmanned Aviation: A brief history of unmanned air vehicles*. AAIA, Baltimore.

6) "World War II: America First Attack Drones"

Spark, N.T. (2011, June). Command Break by Nick Spark (http://bit.ly/1HGU0oY)

7) "The Little Helicopter That Could"

Gyrodyne Helicopter Historical Foundation (http://bit.ly/1LPdxib)

8) The Navy failed DASH. (n.d.) (http://bit.ly/1LPdxib)

9) "The Fighting Fire Flies"

Wagner, W. (1982). *Lightning bugs and other reconnaissance drones*. Fallbrook, CA: Aero Publishers Inc.

10) "The Aquila Fiasco"

Munson, K. (1982). *World unmanned aircraft*. London: Janes Information Group.

Connell, J. (1986). *The new Maginot Line*. London Coronet Books.

11) Fu-Go Windship Weapons

Mikeshi, R.C. (1973). *Japan's World War II Balloon Bomb Attacks on North America.* Washington: Smithsonian Institution Scholarly Press.

Webber, B. (1992). *Silent Siege III: Japanese attacks on North America in WWII.* Central Point, OR: Webb Research Group.

CHAPTER TWO

THE SURPRISING RISE AND SUDDEN FALL OF THE PREDATOR

> *"Sasha and Malia are huge fans, but boys, don't get any ideas. Two words for you: Predator drones."*
>
> - President Barack Obama to the Jonas Brothers, White House dinner 2010

By 2010, drone strikes had become such a part of national consciousness that the President could joke, rather tastelessly, about using them to keep unwelcome suitors away from his daughters. It is easy to forget the long and difficult path that led to this casual acceptance.

In the 1980s drones were in the doldrums. Arnold Schwarzenegger might have been rampaging around as Hollywood's Terminator, but in the real world, combat robots were nonexistent. There was simply no demand for unmanned aircraft in the military.

The electronics revolution was in full swing. Digital technology made flight electronics flexible and reliable, the new Global Positioning System was well under way, and satellite communications meant that you could stay in touch anywhere on the surface of the Earth. The pieces were falling into place for a drone far more capable than its predecessors, a drone that would change everything. It was called Predator.

As an aircraft, the Predator is decidedly unimpressive. Its performance in the air is poor compared to many World War I biplanes from a century ago. It is frequently derided as a "plastic airplane" or giant model aircraft, even by those who fly it. The pilot's view is so restricted that flying one is compared to looking at the world through a drinking straw.

Few would call the Predator a design classic; it is more a technological kludge of different components tacked together, with an engine derived from a snowmobile at the back, an outsize satellite communications pod stuck on top, and missiles so heavy it can barely carry them slung underneath. According to one estimate, it takes seventeen people just to fly this "unmanned" aircraft. And yet the General Atomics MQ-1 Predator has been immeasurably more successful than any previous drone.

Origins

The Predator did not have an easy start, and might easily have fallen victim to the type of problems that terminated so many earlier projects. DARPA, the Pentagon's Defense Advanced Research Projects Agency, identified the need for a long-endurance drone and had carried out a classified study under the codename Teal Rain in the 1970s. This led to the construction of an aircraft called Amber (1) with a wooden propeller and a distinctive upside-down-V tail.

Amber was designed by Abraham Karem, an expert on gliders and other soaring aircraft. Ten examples were built, and performed well in tests, but the project was wiped out by budget cuts anyway. Karem, despite his reputation for brilliance, was too prickly and outspoken to be an effective advocate for his drone. Leading Systems, the company which made Amber, promptly went out of business. That might have been the end of the story.

The company's assets were bought up by General Atomics, an outfit originally set up to harness nuclear power but now diversifying into other technologies. Leading Systems' most important asset was a cheap export version of the Amber called the GNAT-750. The Turkish government had expressed an interest in the GNAT-750, a larger version of Amber, with a wingspan of thirty-six feet and an empty weight of five hundred pounds. Being an export model, it had less expensive electronics. The engine was a German Rotax 914 used in sailplanes and light aircraft (a smaller version is used in snowmobiles).

The GNAT-750 flew at barely a hundred miles an hour, but resembling a glider it required minimal power to stay in the air. A flight endurance of around forty-eight hours meant the GNAT-750 could maintain constant watch over a given area for longer than any manned aircraft.

However, the deal went bad: the Turkish government was unable to pay for the drones it had ordered. Once again, the project appeared to have reached a dead end.

When the 1993 conflict in Bosnia flared up, the US had no suitable reconnaissance drones on hand. Satellites were unable to see beneath the cloud cover. Existing spy planes were designed to operate in hostile skies, flying at extreme altitude like the U-2 or at extreme speed like the Mach-3 SR-71 Blackbird. The requirement was for a drone that could fly at low speed and low altitude, carry an off-the-shelf camera system, and beam back real-time video via a relay aircraft. CIA Director James Woolsey had worked with Karem on a project related to the MX missile and was impressed with his expertise and creativity. The GNAT-750 looked like the ideal solution. It provided a stable platform with long endurance and, because it was "export technology," there was nothing sensitive that would cause problems if one was shot down and the remains analyzed.

All the GNAT-750 needed was the communications link to a relay aircraft. The relay aircraft was the link between the drone and the controller on the ground; by maintaining line-of-sight with both the drone and the ground station, the relay plane could keep the two connected. The rapid development program was handled by the CIA, an organization with some background in dealing with spy drones.

Even then an accident almost ended flight testing. One of the modifications overseen by the CIA was a security feature that shut down everything if the speed dropped too low, as it was assumed the aircraft must be on the ground. A gust of wind from behind caused the flight speed indicator to drop below the vital figure. The GNAT-750 duly switched itself off and dropped like a stone.

That sort of accident could kill a manned program along with the pilot, but the loss of a drone is not such a serious matter. The program was completed in months. The CIA operated the GNAT-

750 from Albania, flying missions to Bosnia with considerable success. Video was sent back via the manned relay aircraft – like the earlier TDR-1, DASH, and Firebee, radio range was the limitation – and missions only lasted as long as the relay plane was in place.

The GNAT-750 flew slowly and sounded like a lawnmower, but it sent back good clear images, sharp enough to identify military vehicles or distinguish surface-to-air missile sites from decoys. The resolution on the ground was eighteen inches – as good as many satellites, with the advantage that it could be sent when and where needed, whereas satellites only appear every ninety minutes as their orbit allows. There were problems – the communications relay was unreliable and kept dropping – but the drone returned good intelligence for a modest outlay.

The drone turned out to be stealthy, not from design but because it was largely made of composite material and there was not much metal to give a radar return. The GNAT-750 might have been useless in a major war against an enemy with effective anti-aircraft defenses and fighter jets. It lacked all the fancy bells and whistles that Army planners had envisaged for Aquila. But in this situation, the humble GNAT-750 was ideal. All four aircraft originally slated for delivery to Turkey ended up being purchased by the CIA.

The Pentagon was not content to let the CIA have a monopoly on drones. As it was apparent that there might be further limited conflicts where such drones could be useful, they funded their own development of the GNAT-750. This was an Advanced Concept Technology Demonstration or ACTD for a version known as the 750-45 or 750-TE Predator. The Predator name was chosen after a competition among General Atomics employees.

The result was a larger aircraft; the empty weight almost doubled to a thousand pounds. It could stay in position five hundred miles from its base for twenty-four hours. Most important, it had extra communication equipment, including a large and unwieldy but effective Ku-band satellite communications setup with a gimbaled antenna that swivels around under its cover to keep pointing at a satellite. Sudden maneuvers tended to break the link and contact could be lost for a minute; the autopilot kicked in while the drone found its satellite again. While it was not be entirely reliable, armed with this capability, the new drone could beam back video from anywhere in the world without a relay plane. And it could fly anywhere, watching for as long as fuel lasted. It entered service in 1995 as the RQ-1 Predator.

The Predator ACTD project was a triumph for the Air Force's BIG SAFARI team. Technology alone does not succeed without an organization dedicated to cutting through red tape and making things happen fast.

The Predator saw action immediately in the Bosnian conflict; one was shot down by a Serb anti-aircraft unit. A few months earlier, USAF pilot Scott O'Grady was shot down by the same unit in the same area, an event that brought massive media coverage and a huge rescue operation. O'Grady was eventually picked up by a rescue force comprising two rescue helicopters, two helicopter gunship escorts, two Harrier attack aircraft, four electronic jamming aircraft, a couple of fighters providing high-altitude cover, and some tank-busting A-10s for good measure. No such response was needed when the Predator went down. There would be no *Behind Enemy Lines* – Hollywood's version of O'Grady's narrow escape – when a drone was lost. Sometimes it is better not to have a man in the cockpit.

In the subsequent Kosovo campaign, the Predator watched over columns of refugees and assessed battle damage. At ten thousand feet it was inaudible, and rarely noticed by those on the ground unless they actually craned their head back to look at it. Skilled operators learned to use the cover of the sun to shield their aircraft from those they were watching.

In former Yugoslavia the Predator was of little use in directing air strikes due to a lack of training and poor communication between different units. One officer complained it took about forty-five minutes to get a strike aircraft into the same area as the drone, while the drone operators sometimes provided poor descriptions of the target – "the house with orange tiles" was not enough in a village with twenty of them. This experience prompted the addition of a laser illuminator to the Predator so the operator could highlight an aim point by shining the laser light on it, "sparkling" it, in in Air Force slang. The laser indicated the target for pilots far more quickly and easily than words.

In a later addition, originally known as Wartime Integrated Laser Designator (WILD) (2), Predators were fitted with lasers to mark targets so laser-guided weapons could home in on them – "lasing" rather than just "sparkling." A pilot releasing a weapon did not even have to see the target so long as the Predator operator could keep it lit up.

Some believed that the Air Force would lose interest in the Predator after the conflict in former Yugoslavia. Just as the end of WWII and Vietnam had seen drones shelved because they were no longer needed, the Predator looked like it might end up back in a warehouse or assigned to target duties. Having survived two dead ends, the third one would probably have been final. Then there were the terrorist attacks of 9/11 and everything changed.

The 2001 invasion of Afghanistan and the 2003 war in Iraq both turned into counterinsurgency operations. There was no air defense to worry about, but there was an urgent need for reconnaissance aircraft that could watch over vast swathes of countryside for long periods.

Even then, it was not guaranteed that Predator would survive. Many critics disliked it. A 2001 report noted that the drone was unreliable, and that rain in particular affected the Predator's ability to locate targets or indeed see anything at all on the ground.

Losses were high. By 2001, twenty of the sixty Predators had been lost to a mixture of pilot error, bad weather, accidents, and enemy fire. "Situational awareness" in unmanned aircraft is notoriously poor because of the limited view and the lack of feedback from other senses. You cannot hear the engine or feel vibration. The extreme case occurred when a pilot crashed during landing because she did not realize that her Predator had been flipped over and was flying upside down. Lesser mishaps are common. The accident rate peaked at one crash per 2,500 hours flown, far higher than any manned aircraft – but not unusual for a drone.

The loss rate was considered to be acceptable. At less than $3 million an airframe, compared to over $200 million for some manned jets, and with no pilot casualties to worry about, the Predator was expendable. Improvements in training and additional safety features brought the accident rate down to one per 20,000 hours in 2010. By 2013, large drones had a lower accident rate than many manned aircraft.

In spite of doubts over the loss rate, the Predator was a winner. Its trump card was its unique ability to beam back real-time images while patrolling in endless circles.

"This aircraft is known for its video," an Air Force official stated at a press briefing in 2001. "It has become the commander's real-time eye in the sky, providing real-time streaming video back to the command post."

In addition, the video can be transmitted to any other interested viewers – including the Commander-in-Chief if necessary. The imagery appeared to be highly addictive, leading to it being called "Predator crack" because it seemed that commanders right up to the White House could never get enough of it. The Predator provides a front-row seat to the action that was previously unavailable to commanders. For the first time, they could break through levels of subordinates and see for themselves what was happening on the battlefield, when it was happening.

On the other hand, intelligence analysts were accustomed to imagery in the form of black-and-white still photographs, not color video. Initially their approach was to take stills from the video feed and print them out. It might seem bizarre, but this was a perfectly rational approach for people whose skills lay in interpreting photographs rather than video. Later, the advantages of the video feed became apparent and soon the analysts were taking full advantage of it.

The drone had finally found its niche. It thrived and multiplied.

Predator Operations

The USAF developed a whole unmanned aircraft operations methodology around the Predator. The plane can be broken down and stored in a shipping container known as a "coffin" and flown around the world on a transport aircraft. In theatre, it needs a five-thousand foot runway and a dedicated ground support team. Once it is rolled onto the runway, the drone is piloted by a local crew who get it into the air and on its way. Then it is handed over via satellite link – it has its own special twenty-foot dish and dedicated satellite systems – to a remote team. From then on the Predator is flown from Creech Air Force Base, forty minutes outside Las Vegas.

Unmanned aircraft like Predator have major support requirements, but each flight provides twenty-four hours of continuous surveillance on station with cameras, infrared, and radar sensors. By using the drones in relays, the Air Force can maintain a permanent presence over an area, known as a "combat air patrol," "CAP" or "orbit." Each CAP requires at least three drones. In 2010 there were fifty Predator/Reaper CAPs; by 2013 the number was up to sixty-five, with plans to replace all the Predators with Reapers by 2016.

The drone operator, sometimes called the pilot, works at a Ground Control Station or GCS with two twenty-inch monitor screens. The upper screen shows a map of the area where the Predator is flying, the lower screen shows the view from one of the nose cameras as well as the instrument display. The pilot flies with the aid of a conventional control stick, throttle, and rudder pedals, just like other aircraft. The display screen also has a section resembling an online chat room used to communicate with troops, intelligence officers, and pilots in the theatre of operations.

Initially the pilots were recruited from existing combat pilots who had flown other aircraft, but later the Air Force started training pilots specially. Predator operations take a lot of piloting. The plane may keep going for twenty-four hours, but that requires several shifts of pilots, with replacements for those that are sick or otherwise not available. You need pilots in theatre to land and take off, as well as those at Creech. One study suggested that ten pilots were needed for each predator CAP to keep operations going 24/7. These days less than half of drone pilots qualified on other aircraft first. Pure drone pilots may have some advantages; reports suggests that pilots have to unlearn some of their skills before they can fly the Predator effectively, as they may have become reliant on feeling the tilt of the aircraft or the change in note of the engine to tell how it was flying.

But as far as the operators are concerned, the experience is just like being there.

"Physically, we may be in Vegas, but mentally we're flying over Iraq. It feels real," says one drone pilot.

One of the biggest differences from other aircraft is the time lag of a few seconds (latency) due to the satellite communications. However rapidly you pull the stick, the Predator cannot respond until two seconds later. This latency of a few seconds means the pilot cannot react quickly to threats, coupled with the lack of situational awareness, makes them vulnerable to attack from the air or the ground.

The Predator has a nose camera, giving the pilot the equivalent of the view from the cockpit – albeit a very restricted view compared to most aircraft. The real business end of the Predator is a "sensor ball" eighteen inches in diameter. This is the AN/AAS-52 Multi-Spectral Targeting System (3), which has a stabilized gimbal mount with two axes of rotation, keeping the cameras pointed in exactly the same direction regardless of the motion of the drone. It has normal visible-light cameras for daytime use and image-intensified night cameras, as well as infrared imaging, along with software that combines the inputs from different cameras into a single image. It features various levels of zoom, from a forty-five degree wide-angle view down to an ultra-narrow 0.2-degree view. This is equal to a x200 zoom range. On a standard 35mm camera, the equivalent lenses at the extreme ends would be a 50mm wide-angle lens and a 12,000mm telephoto. Even from ten thousand feet, the Predator's lenses can keep a close eye on unwitting subjects below.

The sensor ball also contains a laser illuminator, like an invisible searchlight indicating targets for friendly forces, a laser designator for the Hellfire missile, and a laser rangefinder to determine the exact location of the target.

Even in pitch darkness – or in the rain, which was a problem previously – the Predator can pick out objects on the ground with great accuracy. This is thanks to a radar system called Lynx developed in 1998 by Sandia National Laboratories (4) to overcome the limitations of cameras. Existing radar was too large for the Predator at some four hundred pounds. In a major feat of miniaturization, the necessary electronics were crammed into a package weighing just a hundred and twenty pounds which generates an image resembling a black and white video with an impressive level of detail.

From fifteen miles away, Lynx produces images in which features four inches across can be distinguished. It also has some other clever tricks. A process called coherent change detection shows the difference between the current scene and one recorded previously. This is accurate enough to pick up the disturbance left by a bomb buried under the road surface. Interestingly, the radar has reportedly been omitted from some of the later Predators as it was not found necessary for their missions. It may be that in fine weather the radar is surplus to requirements.

The Predator can also carry various electronic warfare packages that allow it to detect, locate, and intercept radio signals. The simplest of these was a radio receiver bought from Radio Shack; the most advanced are highly classified and cost millions. These could, for example, pick up walkie-talkie or cell phone transmissions and pinpoint the users. Predators can reportedly track individual cell phones when they are on by their SIM cards. Intelligence analysis may be able to identify the language or even the identity of the speaker. They may also be able to detect radars, mapping the location and type of every transmitter in range.

In addition to the pilot, there is at least one, and often two or three sensor operators with similar displays. While the pilot is a flight officer, the "payload handlers" are usually enlisted men or women. Their role is to monitor what the sensor systems are picking up, and ensure that the cameras are pointed at items of interest. Unlike pilots, they are intelligence specialists with no flight experience. They are backed up by a further team of analysts – often a dozen or more for one mission – who look at the camera and other feeds in greater detail after the event. They may be looking to identify particular individuals or vehicles, to spot patterns of behavior, or anything else of interest. During counterinsurgency operations this might be a delivery of groceries to a supposedly unoccupied house or an unusual gathering of pickup trucks at a spot in the desert. In support operations, such as in Libya, they may be looking at troop formations or defense works. Strategically, they might be looking for signs of the infrastructure that goes with chemical, biological, or nuclear weapons production.

The overall number of people involved in maintaining a single CAP may be as many as two hundred, including maintenance, launch and recovery, flying, and the extensive teams for PED – "processing, exploitation and dissemination" of the information gleaned.

Drones can be highly effective in counterinsurgency. Perhaps their biggest single success has been Task Force ODIN (5), a US Army unit tasked with tackling the problem of roadside bombs in Iraq. ODIN stands for Observe, Detect, Identify, and Neutralize, and the unit flew a version of the Predator called Grey Eagle. A variety of sensors were used to spot insurgents planting bombs. These would be followed back to other insurgents so the entire insurgent network could be mapped out before strikes and ground raids were sent in. ODIN operated some twenty-six drones in conjunction with manned aircraft and ground units. The unit is credited with killing three thousand insurgents in one year and turning the tide against IEDs.

ODIN operators were known for their skill at assessing the situation on the ground and getting to know the pattern of life in a village. They were trained to recognize the features that distinguished fortified compounds from ordinary homesteads, and the tell-tales signs of roadside bomb emplacement. They claimed to be able to tell whether a group of men on the street were just local militia or whether they were a threat to US forces. It is hard to assess how many of their targets were valid and whether ODIN really was the determining factor in the reduction in insurgent activity. But the undoubted technical skill of ODIN, and their ability to "find, fix, and finish" those planting IEDs certainly played a part.

The soldiers on the ground seeing Predators in their area knew that high-magnification video cameras were being trained on their targets, but they were completely out of the loop. Until, that is, the BIG SAFARI team came up with a device called the Remote Operations (or Remotely Operated) Video Enhanced Receiver or ROVER (6). This was a simple video receiver for the ground troops; they couldn't control the Predator, but they could see what it was seeing, and the information was invaluable. The initial version was so big it had to be carried in a Humvee, but later version have been laptop sized and smaller, looking like handheld gaming consoles. Later versions have also included interactive features so the user can contact the Predator pilot to make requests with the "chat room" function.

ROVER has become a key tool for special operations and others. In 2006, USAF Lt Col Wayne Shaw wrote, "The one question on the minds of ground brigade commanders was 'How many ROVERs were available and were there extras?'"

The drones can work in friendly airspace, like Afghanistan. They also operate in grey areas over Somalia, Yemen, and Pakistan where the locals are likely to be unfriendly. They have also carried out reconnaissance (though not strike) missions elsewhere. At least two have crashed inside Iran.

In spite of how valuable their intelligence is, the Air Force does not give Predator pilots much help. The remote nature of operations means that they lead a surreal existence. They spend hours watching over compounds and roads in places thousands of miles away. Drone operators often quote variations of the line about warfare being "long periods of boredom punctuated by moments of extreme violence"; they may be called on to kill, and often see the aftermath of violence in great detail for prolonged periods. This may include watching bloodied victims attempting to crawl away from a strike, or attempting to count bodies that have been blown apart.

At the end of a shift they hand the drone over to another operator and go back home to their family. This may be remote warfare, but a 2008 study found drone operators have high levels of "fatigue, emotional exhaustion and burnout" compared to traditional aircrew flying in the same combat zones.

Nor are drone pilots much appreciated inside the military. Studies suggest that drone pilots are less likely to be promoted to senior positions compared to those who sit in a cockpit. Pay for drone pilots is now on a par with others, but other pilots are eligible for "continuation pay" of up to $25,000 a year (7). A looming shortage of drone pilots led to this policy being reconsidered in 2015.

In 2013 the Pentagon abandoned plans for a "Distinguished Warfare Medal" that would be awarded to drone pilots who had an impact on combat operations while not actually serving on the front line. Congress had attacked the new medal, which would rank higher than a Bronze Star or Purple Heart, as "a disservice to our service members and veterans who have, or who currently are, serving overseas in hostile and austere conditions."

Drone operations are clearly not ranked the same as manned ones.

Armed Robots

In 2000 a CIA Predator drone spotted a tall figure in flowing white robes and surrounded by bodyguards at an al Qaeda training camp called Tarnak Farms in Afghanistan. Analysts quickly identified the man as Osama bin Laden, already a wanted terrorist after the bomb attack on the World Trade Center in 1993 (8). However, the only way of attacking him would be to call in a cruise missile strike. A cruise missile is a blunt instrument: with a warhead of a thousand pounds of explosives, it causes destruction over a wide area. Previous missions had noted the presence of children's swings at Tarnak Farms, and President Clinton was said to be concerned about the level of civilian casualties. In addition, cruise missiles would take several hours to arrive, by which time bin Laden would probably be long gone. There was no strike.

The same problem occurred numerous times during Operation Allied Force in Kosovo. Predators would locate a target, but it would be gone before attack aircraft could be sent in. The obvious way to reduce the "sensor to shooter" time, the delay between seeing a target from a Predator and attacking, was to put a missile on board. Cofer Black, head of the CIA's counterterrorist Center, requested that the Predator be armed. USAF General John Jumper requested a demonstration of a Predator that could "find a target, then eliminate it."

Previous work indicated that it should be possible to fire a Hellfire from the Predator, although this was far from certain. It is a big missile for a small plane. Hellfire was chosen because it was a proven, mature missile with many years of successful service. In fact, mature is something of an understatement: Hellfire was positively middle-aged, dating back to the Nixon administration in 1974.

When Hellfire was designed, the major concern was an onslaught of Soviet heavy armor in Europe. It was intended to be delivered by a helicopter; the contrived name is short for "HELicopter Launched FIRE-and-forget missile." The "fire and forget" was because a TV-based guidance system would lock on to the target and do the rest. However, the video guidance was dropped when it was found to be unsatisfactory. Ever since the TDR-1 there had been a series of TV-guided missiles and bombs, but they suffered the same problems with low-visibility targets.

Hellfire has laser guidance, so the target needs to be marked with a laser designator right up until the missile hits. It is highly accurate, usually hitting within half a meter of the aim point, but it can take a while to arrive. Hellfire is faster than the speed of sound, but fired from a range of six miles, it still takes about twenty seconds to reach the target. That twenty seconds can feel like an eternity, especially in a situation where the target may disappear, or more people or vehicles might appear within the Hellfire's lethal radius.

Hellfire is uncomfortably large to hang off a Predator but small compared to anything else available in the inventory. At the time, the Air Force's smallest guided bomb was five hundred pounds. The Hellfire is an Army weapon not Air Force, and there were doubts about whether it could be successfully integrated, but there were no good alternatives. In February 2001 the first test firing was carried out. Success was by no means certain, with some developers worried that a missile launch would tear the Predator's wing off, or that the exhaust would scorch the drone or blind its cameras. The wings were strengthened with additional bracing.

The trial was a success, and a laser-guided missile with an inert payload instead of high explosives made "a nice dent" in the turret of the tank target. There was some damage to the Predator.

"We also dissembled a wing to see how much fracturing we had," said an Air Force official after the event, "and we didn't have that much." (9)

Further strengthening was added, and the armed Predator went into service.

A few months later a Predator launched its first Hellfire in Afghanistan at a group of suspected insurgents. Sixty-five years after the TDR-1 assault drone, forty years after the QH-50 DASH gunship, and thirty years after the Model 234 Firefly, the US once again had an armed flying robot at its service.

In the early days, launching a Hellfire was not a simple process.

"I have to make seventeen-plus different mouse clicks in pull-down menus," complained one drone pilot to National Defense Magazine. "In my aviator thought-process, that doesn't make much sense. I would much rather have a cockpit where I reach over here, arm, select and shoot."

The pilot noted that in his previous aircraft, an F-16, weapons release required a single button push.

The other challenge when firing a Hellfire from a Predator is the unavoidable two-second time lag caused by satellite communications. This means the laser spot effectively takes two seconds to move, so the Predator can only engage a stationary target in this mode. To hit a moving target, the operator uses targeting software to lock on to a moving vehicle; the software keeps the laser spot in place, an indirect way of engaging the target with its own risks if the system fails.

The operator witnesses the missile strike as a whiteout on the screen, followed by pixelation and the bloom of the explosion. Then they get to see the effects of the strike in close-up detail. Hellfire does not tidily disintegrate the target as in a video game but leaves recognizable bodies and body parts around ground zero. People running away from the strike are known as "squirters" – either because they squirt out of the strike area in all directions, or from the way they shoot randomly, depending on who you believe.

Predators are often used for post-strike "battle damage assessment" to gauge the effects on the ground. Looking at the effects of your own strike requires a strong nerve and a belief in the principles of remote warfare.

There are several versions of the Hellfire missile in use. The AGM-114K is the basic anti-tank weapon with a shaped-charge warhead. This focuses the explosive power into crumpling a metal cone and converting it into a high-speed jet of semiliquid metal than can punch through two feet of armor.

The K is highly effective against tanks but less effective against other targets, because so much of the force goes into the armor-piercing jet. A new version was developed with a new metal sleeve around the warhead. The sleeve is blasted into shrapnel, high-velocity metal fragments with lethal effects over a wide radius. Perhaps inevitably, the shrapnel-producing version is called "Special K."

The other main model was the AGM-114M, with a warhead that combines blast and fragmentation. This is fitted with a delayed-action fuse so it will go through a brick wall, roof, or other obstacle before detonating inside a structure or vehicle. It is highly effective against targets inside buildings.

The AGM-114N has a "metal augmented charge" or thermobaric warhead. This is optimized to produce the maximum amount of blast and can completely level most buildings. We will look at thermobarics in more detail in Chapter 8.

Later on two new versions were produced, the AGM-114P and the latest AGM-114R version. These added trajectory shaping: rather than flying along the shortest route to a target, the missile can dive down from above, useful when fighting in city streets. The R version has an inertial sensor that communicates with the launch vehicle. It can be fired in any direction and then turn around and head towards where the laser is pointing. The aircraft does not need to face the target to fire, saving valuable time when the target may only be visible for seconds.

If fired in a straight line, Hellfire's supersonic speed means the target will not hear it coming. The majority of Predator victims aren't even aware they are being targeted until the missile hits. However, in some trajectories, or if the missile has to turn and go back on itself, there will be an audible sonic boom before it hits.

Predator operators have reported that dogs often notice the missile coming and get out of the target area, possibly because of an ultrasonic sound, but this is purely anecdotal. Insurgents know to move fast when they hear the boom, though running may be less effective than lying flat. Those who simply lie down sometimes survive the strike.

"You might just blow out their eardrums," claimed one drone instructor. (10)

Hellfire may be precise, but it is not surgical. The twenty-pound high-explosive warhead can cause major "collateral damage," killing innocent bystanders or building occupants when the target is a single terrorist.

Worse, the long time of flight means there is the risk of somebody wandering into the target area after the missile has been launched. In his book *Predator: The Remote Control Air War Over Iraq and Afghanistan*, pilot Matt Martin describes an attack where the target was a parked truck with insurgents loitering around it. A missile was launched at the truck, and then he saw two boys aged about ten approaching on a bicycle. There was nothing he could do; diverting the missile might have caused even worse damage. So Martin and his colleagues watched the missile hit, and afterwards saw the dead bodies of the two boys among those of the insurgents.(11)

Armed Predators have been used for a complete range of combat roles, including close support of troops in combat, something that was never envisaged initially. In 2005, a Predator was diverted from its mission and sent to assist a unit of US Marines under mortar attack. The Predator operator destroyed the mortar position with a Hellfire. In similar incidents, Predators were used to direct air strikes or designate targets for guided bombs, or take out ground targets on their own.

The most celebrated case of close support was during Operation Anaconda, when a Predator was sent to assist US troops who had survived a helicopter crash and were under heavy attack from the Taliban. For eleven hours the Predator circled overhead, lasing targets for US Navy planes to bomb. When rescue helicopters finally arrived, the Predator's laser illuminator directed them to a safe landing site.

However, it can be difficult to work together with someone physically thousands of miles away without errors. In 2011 the first drone friendly-fire incident occurred when two Marines were killed in a strike in Afghanistan. They were moving to reinforce a Marine position under attack when drone operators mistook them for part of the attacking force with fatal consequences. (12)

Predators have also been used for perimeter defense around their own air bases. When a Predator arrived back at base with flying time to spare, the local crew used it to patrol the area for insurgents preparing mortar or rocket attacks. The Predator is highly effective at this sort of "force protection" mission, though it is only necessary because the Predator and its ground crew have to be located close to the theatre of operations.

Targeted Killing

Other combat missions are secondary to the armed Predator's main strategic role, strikes on "high value targets," individuals identified as terrorists or insurgents. These strikes are sometimes called "targeted killings" and are often in remote areas in Pakistan, Afghanistan or Yemen. Although much of the intelligence gathering may be carried out via the Predators themselves, operators also rely on what is known as HUMINT, "human intelligence," old-fashioned spies on the ground.

When tracking insurgents planting bombs or carrying out mortar attacks, the eye in the sky can see everything. But when you are looking for a specific individual, it takes a more personal approach. Insurgents may be tracked and targeted by their mobile phones; it has been suggested that more missiles have been fired at known SIM cards than at recognised individuals. In recent years agents in the field have marked targets for drone strikes with miniature radio beacons. A mobile phone left under a seat can perform the same function. The Taliban in Afghanistan have executed alleged spies who left electronic devices concealed in empty cigarette packets. Taliban commanders now know to leave their vehicles guarded at all times to prevent tracking devices from being attached.

Much of the technology in this area is classified, but we do know that the Pentagon has devoted considerable effort to tagging and tracking technology, much of it devoted to marking individuals. Radio beacons and transponders which require a power source are comparatively bulky due to the size of the batteries. Other types of tagging are passive. One approach is to shower the target with a fine dust of "quantum dots", tiny specks of semiconducting crystal. These are invisible to the naked eye but can be detected from long range with the aid of an infrared laser illuminator. Different batches of dots can be given specific codes, so a tagged individual or vehicle can be identified days later from long range.

Other tagging technologies include dyes and inks visible only through special viewers. One DARPA document even suggested that an additive could be introduced to the target's shampoo so they could be identified and tracked. It is not clear whether this idea was ever developed.(13)

Drone strikes have continued in Afghanistan, Pakistan, and Yemen. Such strikes are often extremely controversial. An observer monitoring a sensor screen at Creech Air Force Base may see things very different to people on the ground.

For example, in December 2013 a US drone launched four missiles at a vehicle convoy in Yemen, killing twelve people and injuring many more. The authorities describe the targets as terrorists belonging to Al-Qaeda in the Arabian Peninsula (AQAP). Witnesses and local people interviewed afterwards said the convoy was a wedding procession, travelling from a wedding lunch to a second celebration at the groom's village, and that one of the injured was the bride. They say the victims had no connection with terrorism and thought nothing of it when they heard the drone overhead. Then missiles started coming down:

"Blood was everywhere, the bodies of the people who were killed and injured were scattered everywhere.... I saw the missile hit the car that was just behind the car driven by my son..." (14)

The angry relatives displayed the bodies of the dead at a nearby town. They were paid compensation by the local government, suggesting their claims were accepted. The drone performed perfectly, but the intelligence that guided the strike was faulty.

The dangers of faulty intelligence were highlighted during the 'tall man' incident of 2002. Osama bin Laden was notably tall at six feet five; when CIA Predator operators saw a very tall man apparently surrounded by bodyguards at a site near Khost, a Hellfire strike was authorized. The tall man was killed, but it turned out that he and his companions were harmless scrap metal collectors. The intelligence analysts had completely misread the situation in their enthusiasm to get the al Qaeda leader.

In the West, there has been concern over "video game war" and the alienating effect of distance. It seems easier to kill a figure on a screen than a living, breathing human. The military insist that unmanned strikes are conducted according to the same standards as manned ones. But the fact that drone operators describe those on the receiving end as "bugsplat" has infuriated critics.

The term *bugsplat* was actually the informal name of a computer program to calculate the effects of a missile strike and the likely amount of collateral damage. (15) The programmers were a long way from the battlefield, but the jokey name continued in use. The software is correctly known as FAST-CD, or Fast Assessment Strike Tool — Collateral Damage.

The number of civilians killed by drone strikes is a matter of hot debate. An investigation by the New York Times found that in official reports, all military-age males killed in strikes were considered to be combatants "unless there is explicit intelligence posthumously proving them innocent." Estimates from 2014 of the number of civilians killed in Pakistan range from less than a hundred and fifty to almost a thousand, out of approximately three thousand killed.

Drone strikes have other effects too. The "Living Under Drones" report by Stanford International Human Rights and Conflict Resolution Clinic suggests that the drone war in Pakistan is having long-term impact on the mental health of those who live with it. They say that the drones cause "constant and severe fear, anxiety, and stress" to people who can do nothing to protect themselves, and who may develop phobias about social gatherings, attending school, or even going to the market as a result of witnessing missile strikes or their aftermath.

The strikes have been highly effective at taking out specific named targets, and there is no doubt that the leadership of Al-Qaeda and other groups has been decimated by the drones. Destroying the leadership does not necessarily bring victory; a leaked 2009 CIA report on targeted killings noted that they had limited effect against the Taliban in Afghanistan because leaders were replaced so rapidly. (16) Others have cast doubt on the use of drone strikes against high-ranking individuals because they reduce the chances of any negotiations with the group.

Whether the strategy of targeted killings can produce victory has been questioned, but the fact remains that drones can reach places that are otherwise inaccessible, and they are uniquely effective at placing a missile on a target. When politicians are under pressure to do something in response to a crisis, a drone strike is a tangible way of showing commitment and striking back at the enemy without risking any friendly casualties.

They may be politically toxic, but it's hard to see leadership relinquishing such a useful tool as drone strikes.

Reaper and the Fall of the Big Drone

As noted, the Hellfire is not an ideal fit for the Predator, which struggles to carry two of them. Rather than developing smaller, smarter weapons, the Air Force decided it wanted a bigger aircraft. General Atomics anticipated this and the company funded development of "Predator B." In October 2001, the Air Force bought two of the new drones for evaluation; in 2004 they ran a competition for a new "hunter-killer" drone, but the results were perhaps a foregone conclusion.

When it went into service in Afghanistan in 2007, the Predator B was renamed the "MQ-9B Hunter-Killer" or "Reaper." The Reaper is a far more substantial aircraft than its flimsy predecessor. It is four times as heavy; the turboprop engine is six times as powerful and doubles to speed to around 200 mph. It looks less like a big toy and more like a traditional combat aircraft, the sort that the military are more comfortable with.

A Reaper can carry fourteen Hellfire missiles, or four missiles and a pair of laser-guided 500-pound bombs. These are useful against bunkers, large buildings and other targets too big for the Hellfire. The selling point of the Reaper is "deadly persistence": it can loiter over an area for a prolonged period and engage a large number of targets of opportunity as they present themselves. This puts it ahead of the Predator, which can only attack two targets at most. However, carrying some three thousand pounds of ordnance cuts the Reaper's endurance to around fourteen hours.

The Reaper has been popular with the Pentagon. The last Predator was bought in 2011, the idea being to gradually replace them with the larger, heavier successor.

The Reaper's increased size comes with an increased cost. The flyaway price for Reaper is around $14 million for the basic model, or $20 million with all the trimmings, compared to $4 million for a Predator. Correspondingly fewer have been acquired. While the Predator has always been regarded as an expendable aircraft with a finite lifetime – the Air Force counted on losing seven a year in normal operations – the Reaper is expected to be durable.

So instead of a cheap, ultra-long endurance, expendable drone, the Reaper resembles a manned aircraft. Predator operator Matt Martin describes the Reaper as "a longer-duration, lightly-armed (and much less survivable) version of the F-16." Without the duration and price advantages, the Reaper comes perilously close to being in competition with the manned jets. As we have seen, this is often a fatal situation for a drone in the Air Force.

The Air Force planned to increase the number of Predator/Reaper CAPs to 65 in 2013 with more slated for later. However, by 2014 the climate had changed in the Air Force and there was talk of "rebalancing" the fleet to fight in contested airspace. The 2015

Defense Budget revised the number of Predator/Reaper CAPs down to 55 by 2019. The 2016 budget called for an increase but it is uncertain whether this will happen. "Peak drone" may already have passed as far as the Predator and Reaper are concerned. The Air Force's Reaper purchases are scheduled to end in 2019.

The US Army has bought 103 of its version of the Predator, the MQ-1C Grey Eagle. It has been useful for protecting convoys and providing overwatch and long-range patrolling, and appears popular with those on the ground. The Air Force treats Predator like manned aircraft, and manual landings are the rule. These landings can be tricky, as the Predator's large wing area makes it sensitive to sudden gusts of wind. The Army's Grey Eagle has an automated take-off and landing system, and lower accident rates are expected.

But again, there are questions about how useful it will be in a contested environment, and the budget squeeze is on. Combat divisions would originally get twelve Grey Eagles each. That has now been scaled back to nine, and the US Army is not planning to buy any more after 2016.

The makers, General Atomics, are realistic about their prospects. They have developed an unarmed version of the Predator for the export market, and are now touting the Predator B/Reaper as "the pick-up truck of the future" for carrying large sensor payloads at an affordable price. They also developed a stealthy, third-generation Predator known as the Avenger, but have only sold four so far to the US military.

While there are those who believe in drones – and the Department of Defense's "Unmanned Systems Integrated Roadmap" lays out a future where they will play an increasing role in operations over the next two decades – they are losing the battle for actual funding.

The USAF is pushing back against drones, and unmanned systems are bearing a disproportionate share of the cuts. In 2014 analysts suggested that while the overall Pentagon procurement budget has dropped by around 15%, spending on drones was down more than 25%. This fall-off may accelerate as the Air Force looks to future operations that may bear little resemblance to Iraq or Afghanistan, with particular concern about what they call "near peers" who cannot be beaten as quickly or easily.

USAF General Mike Hostage is a particular critic of the drones and their prospects in future wars.

"Predators and Reapers are useless in a contested environment," he told an audience at the Air Force Association's annual conference in 2013. His position was that they were too vulnerable against modern air defenses. "Pick the smallest, weakest country with the most minimal air force – [it] can deal with a Predator." (17)

Hostage has also criticized the Predator's inability to distinguish targets compared to a pilot in the cockpit. This is odd given that the drone is valued exactly for the clear view from its low-speed, low-level stabilized cameras.

Lt General Robert Otto echoed Hostage's comments, suggesting that stealthy manned spy planes should replace the drones after the end of the Afghanistan deployment.

This view is reflected in the effective cancellation in 2012 of the MQ-X program, a next-generation attack drone that would have succeeded Predator and Reaper. There is no reason why drones cannot be stealthy, agile, high-speed fighters. The problem is that this brings them directly into competition with manned aircraft and, as we have seen, any drone that threatens a manned aircraft program is unlikely to survive.

The Last Combat Drone?

The Air Force did once have plans for a fast, stealthy combat drone under a joint project with the Navy called J-UCAS. But the Air Force pulled out of the project in 2006, and after that the Navy program progressed slowly as the experimental X-47 carrier-borne strike drone. This was originally intended for the same sort of missions as manned carrier planes, but commentators have noted how the Navy has been nibbling away at the specifications.

Now known as UCLASS (for Unmanned Carrier-Launched Airborne Surveillance and Strike), it was at first described as a strike aircraft but is now being redefined as a reconnaissance asset for dangerous situations, with some strike capability. It has also been suggested as a flying fuel tank for air-to-air refueling of manned planes. The original J-UCAS should already have been flying; as of 2015, the Navy are talking about fielding in 2022 at the earliest.

According to the Flightglobal website, the specifications for the UCLASS were diluted by the vice-chairman of the Joint Chiefs of Staff, Admiral James Winnefeld. This scaling down has led critics to suggest that there is no role for the UCLASS, and that it is being set up as a sacrifice to the Pentagon budget planners, so that manned programs will remain unscathed. The budget request for UCLASS R&D dropped from over $400m in 2015 to $135m in 2016. This does not look like something the Navy wants badly.

As a combat aircraft, UCLASS looks too expensive to be expendable, and not as capable as manned platforms. History suggests that's a good recipe for termination.

Certainly, there is no sign at present that unmanned aircraft are going to challenge manned ones for supremacy in the eyes of the Pentagon. Their future is still the sleek and glamorous F-35. As the number of Predators and Reapers dwindles, the number of F-35s in service will climb to more than 2,000 over the next two decades on current plans. The pilot's position as knight of the sky is secure.

The Predator changed the world: drone strikes are now part of the lexicon of war. They have enabled the US to take action in corners of the world where manned operations are not feasible. But it looks as though drones have been pushed into a niche role as spies and assassins, leaving the serious fighting to manned aircraft. They might find a new role as transports, but the day of the big fighting drone is over.

However, far below the stratospheric heights of the fighter pilot, there has been a small revolution. Like mammals evolving beneath the feet of lumbering dinosaurs, a very different type of drone has been proliferating close to the ground. These are little craft that do not compete with the lofty lords of the air. And while the big drones are in decline, their miniature cousins have been preparing to inherit the earth.

REFERENCES

1) Project Amber / GNAT 750 –

Peebles, C. (1997). *Dark Eagles: A history of top secret US aircraft programs*. New York, Presidio Press.

2) Wartime Laser Integrated Laser Designator

Grimes, B. (2014). *The History of Big Safari*. Bloomington, IN: Archway Publishing.

3) AN-AAS52 Multi spectral targeting system

Raytheon. (2005). AN/AAS-52 Multi-spectral targeting system. (http://bit.ly/1LRcVs5)

4) Lynx Radar

Sandia National Laboratories. (1999). Lynx: A high-resolution synthetic aperture radar. *SPIE*

Aerosense, 3704, 1-8. (http://1.usa.gov/1N5Lkpc)

5) Task Force ODIN Defense Video & Imagery Distribution System

Bonebrake, C. (n.d.). Angels on our shoulders: Task Force ODIN protects troops on the ground. #.VW8Yo43bK1I (http://bit.ly/1IG7cFF)

6) Rover

Grimes, B. (2014). *The History of Big Safari*. Bloomington, IN: Archway Publishing.

7) Continuation Pay for drone pilots

Irwin, S. I. (2015, January 16). Air Force drone pilot crisis years in the making. [Web log post]. (http://bit.ly/1RpNvsO)

8) Tarnak Farm incident

Pendergast, S. (2014, November 18). Early Predator pilot Swanson discusses his experiences.

[Web log post] (http://bit.ly/1SzOO6R)

9) "We also dissembled a wing…"

Tiron, R. (2001, December). Despite doubts, Air Force stands by Predator. (http://bit.ly/1MYC8oD)

10) "You might just blow out their eardrums"

Mead, C. (2015). A rare look inside the Air Force's drone training classroom - The Atlantic. (http://theatln.tc/UbRkIu)

11) "Martin and his colleagues watched the missile hit"

Martin, M. (2010). *Predator: The remote control air war over Iraq and Afghanistan*. Minneapolis: Zenith Press, 2010.

12) "Friendly Fire" incident

Miklaszewski, J. (Writer). (2011, April 4). 2 US Servicemen mistakenly killed by drone attack in Afghanistan [Television broadcast]. NBC News.

13) Tracking technology, including shampoo Tag, Track, and Locate (TTL) technologies and concepts for combating weapons of mass destruction (CWMD). (n.d.). (http://1.usa.gov/1NJBm2l)

14) "Blood was everywhere, the bodies of the people who were killed"

A Wedding That Became a Funeral. (2014, February 19). (http://bit.ly/21yvaOH)

15) Bugsplat

Washington Post, "'Bugsplat' computer program aims to limit civilian deaths at targets" quoted in *Seattle Times*. (http://bit.ly/1lZGYJB)

16) 2009 CIA report of limited effectiveness on the Taliban

Leaked CIA report: Targeting Taliban leaders 'ineffective' [Television broadcast]. (2009). BBC News.

17) "Predators and Reapers are useless in a contested environment"

Whitlock, C. (2013, November 13). Drone combat missions may be scaled back eventually, Air Force chief says. The Washington Post. (http://wapo.st/1NrZlNd)

CHAPTER THREE

THE RAVEN: A SMALL REVOLUTION

> *The ravens sit on his shoulders and say into his ear all the tidings which they see or hear; they are called thus: Huginn and Muninn. He sends them at day-break to fly about all the world, and they come back at undern-meal [mid-day meal]; thus he is acquainted with many tidings. Therefore men call him Raven-God.*
>
> - "The Tricking of Gylfi," a 13th century Viking epic

Ravens are members of the crow family with a lively intelligence. They will often watch human activity, swooping down to pick up discarded food or roadkill. In captivity, they may even learn to mimic speech. In Norse mythology, the wandering god Odin is always a step ahead of his opponents thanks to his pair of ravens, Huginn and Muninn. The birds fly the world, seeing and hearing everything, returning to perch on Odin's shoulders and tell him the secrets they have discovered.

The most successful drone of all is a personal scout to see ahead and keep you out of danger -- and it is no coincidence that it is called the Raven. Plenty of books have been written about the Predator, but the Raven has largely escaped media attention. It may not look like much compared to its bigger cousins, but the Raven represents a much bigger revolution in drone warfare.

As of 2015 the Pentagon has around ten thousand drones, and nine thousand of them are small, hand-launched craft made by AeroVironment Inc of California (1). The Raven first took off in

2003 and quickly came to dominate the drone world in terms of numbers. It may look like a toy aircraft with a four-foot wingspan, but it puts air power in the hands of the foot soldier. These days, troops may be reluctant to move without their own mini-drones watching over them on foot patrol or flying ahead of their convoy on the road. The Army loves having its own portable air cover, available any time. And in recent years, the Raven has been joined by a miniature strike drone called Switchblade.

Perhaps the real secret of the small drone's success is that it is not seen as an aircraft at all, just as an instrument for seeing over the hill. Big drones compete with the manned aircraft that they resemble, but for once, looking like a toy may be an advantage. Which is why the Raven has become the biggest success story the world of military unmanned aircraft has ever seen.

Beginnings

Paul MacCready grew up in New Haven Connecticut in the 1930s, and was obsessed with flight. As a teenager, he built and flew model planes that broke records. He graduated in physics from Yale in 1947, the same year he established a new soaring record for gliders. He spent the next ten years developing, building, and flying record-breaking gliders. (2)

In 1971, MacCready founded AeroVironment Inc. in Monrovia, California, in the foothills of the San Gabriel Mountains on the northern edge of Los Angeles. The company became famous as the maker of some extraordinary aircraft, based largely on MacCready's work with gliders. In 1977, his pedal-powered Gossamer Condor became the first heavier-than-air craft to fly powered entirely by the pilot's muscular effort. The Gossamer Albatross was the first human-powered craft to cross the English Channel. The Gossamer Penguin and Solar Challenger were solar-powered manned planes, the latter flying from Paris to England in 1981. These aircraft worked because they were efficient gliders, producing substantial amounts of lift from the slightest forward movement.

These successes led to a series of large solar-powered aircraft for NASA and the Pentagon, projects that have so far led nowhere. (We will look at the reasons for this and the problems that face large solar aircraft in Chapter 5). But in 1986 using its own funds,

AeroVironment built a small craft, which they demonstrated to the Marine Corps. The Marines liked it, and the drone went into service as the FQM-151A Pointer. It could be carried on patrol and launched to get real-time streaming video of the view behind the next hill. (4)

The drone had a fixed camera in its nose and had to be pointed at an object to see it, hence the name Pointer. Like the Predator, Pointer was a "pusher" with a propeller at the back of the fuselage so the blades do not obstruct its view.

Pointer was a cumbersome beast, with a wingspan of nine feet and a flight time of just twenty minutes on rechargeable batteries. Its camera had a resolution of just 360 x 380 pixels. But Pointer gave the local commander his very own aircraft, what is known in military circles as an "organic asset," meaning that it belongs to the unit that uses it. Rather than having to go through command channels and talk to the Air Force or an Army reconnaissance helicopter unit, the commander could send a camera over and get a close view of a target in minutes.

Other drones were scarce resources based miles away and available to those further up the chain of command. Mission planning and air traffic control had to be considered. Even if you could get clearance for drone support, it might take hours to arrive, and any intelligence would take even longer to filter back. Pointer could be just snapped together and thrown into the wind. If you wanted to know if there were a dozen enemy tanks on the other side of a ridge, Pointer could tell you immediately.

The Marine Corps took fifty Pointers. They were used during the brief ground action of the 1991 Gulf War, but the mobile tank warfare gave little scope for their use. There was little further interest until 1999 when the US Army ordered four for evaluation. The Army decided that Pointer was too unwieldy and requested a craft with the same capability but half the size, weight, and cost, and with GPS built in. In addition, the new drone would need to be able to carry an infrared camera to see in the dark. Finally, the heavy ground control unit, too big to be truly portable, had to be shrunk.

In 2002, an Advanced Technology Concept Demonstration was launched, the same sort of crash program that turned GNAT into Predator in double-quick time. Known as Pathfinder, this program transformed the unwieldy Pointer into the more compact Flashlite drone. (5) The initial version was difficult to launch and unstable in flight. These faults were corrected with the addition of an airspeed sensor and improved build quality, and over the course of the program, Flashlite became Raven.

The prototype looked good, but General John Keene, Vice Chief of Staff, told the developers that he did not want to wait twenty years before there was a small drone that could be used at platoon level; it had already been seventeen years since Pointer took off. The first test batch of RQ11-A Ravens was delivered to the Army in 2003, just a year after Flashlite.

At this stage the Raven was still handmade as the production volumes were so low. It was a real improvement over the Pointer, being compact, with a wingspan of four feet and weighing 1.9 Kg (4.2 lbs) compared to 4.3 Kg (9.6 lbs) for the earlier version. It was a sturdy construction, made of the same sort of Kevlar and composite materials as combat helmets. The view from the on-board cameras was beamed back to a portable antenna and displayed on the handheld controller from up to six miles away. The new controller was half the size of the Pointer's, and the whole set-up could be easily carried by a two-person team.

Raven's built-in GPS meant it could fly a mission via a series of programmed waypoints with no human intervention, so it could take pictures of a building or installation even if it was out of radio range. Endurance was tripled to an hour, and a new modular design meant changing sensors (say, switching between day cameras and infrared night vision) was a matter of "plug and play". Initially, there were three different nose units: a day camera with forward and sideways-facing camera, an infrared camera looking sideways, or a forwards-looking infrared camera. The sideways-facing cameras mean that the Raven could circle around a target, keeping it in view all the time. The cameras provided four levels of magnification, from wide angle to close-up.

The Raven was an instant hit with troops, and an enthusiastic military press ran stories with headlines like "The Little Plane That Could" and "Rave Reviews for The Raven."(6)

"I now have something as a company commander to give me three-dimensional situational awareness," Colonel John Burke, manager of the US Army's UAV Systems told Army Magazine. "I never had that as an asset of my own before."(7)

Flying the Raven

Raven travels disassembled in a backpack and can be put together in a few minutes. Take-off is achieved by running into the wind and throwing Raven, like a toy glider. Launching is more difficult when there is no wind, and after a bad throw the Raven may glide ten feet before plunging into the ground. Flying in strong winds requires care; winds can reduce the flying time and can cause what are euphemistically called "sudden changes in altitude."

The ground control unit looks like a handheld video game with a joystick and several buttons. The operator can fly manually or simply give the autopilot GPS waypoints. Just tap a point on the map with a stylus, and the Raven flies there. It can be set to orbit a point on the ground, with the sideways-looking camera focused on the centre of the orbit.

Once the mission is over, one stylus tap instructs the Raven to return to its launch point or go to any other location.

Landing is less elegant. When it reaches the landing spot, the Raven pulls up and stalls just above the ground, falling "like a dead bird," according to one operator. A pad on the underside of the fuselage absorbs some of the shock, but the gentle crash-landing looks alarming because the Raven will "disassemble" and come to pieces on landing – "Like a cartoon." (8) The major components are held together with a series of connectors designed to break apart and dissipate the crash energy so the wings, fuselage and payload are not damaged.

Some operators catch the Raven before it hits the ground to reduce the risk of damage. This can become a competitive sport, with operators boasting up to twelve straight catches. However, the makers do not condone catching, as AeroVironment's Steve Gitlin explained to me:

"The perception is that this [catching] will minimize damage, but in reality, it tends to pose a risk to the operator and results in higher damage rates to the system. The aircraft is designed to recover using a deep stall impact to the ground. Other methods of recovery result in loading the structure in ways it was not designed to handle." (9)

The key thing is to be aware that the Raven is coming back. One YouTube video shows an unwary operator struck in the chest by his drone as it returns literally right back to him.

On average, a Raven is reckoned to be able to survive two hundred flights before requiring serious repairs.

"This is dependent on a large number of operational factors," according to Gitlin. "We have common reports of operators achieving 600-800 flights on an air vehicle before requiring repair, and many aircraft still in service are in the 2,000-3,000 flights range."

If something goes wrong, if a battery unexpectedly runs out, the operators may have to go and retrieve their craft. The Army recommends a recovery kit including a folding ladder and climbing spikes for retrievals from trees, plus a supply of trade goods – candy, food, or cash – for bartering with civilians for the return of a missing Raven. It is a rather different operation to retrieving a downed Predator or manned aircraft.

Retrieval may be aided by flashing strobe lights. Some units started putting a label on their Ravens in the local language offering a reward for returning it to the nearest American military unit. These brought back several craft safely, again not something that is feasible with a Predator.

The Raven is powered by rechargeable lithium-ion batteries similar to those in laptop computers. The Army initially tried disposable batteries, but these were rejected because of the hassle of carrying a ton of batteries around. The original battery gave a flight time of about an hour. Changing the battery takes a few seconds, so a Raven returns to its launch point, gets a fresh battery, and flies right out again. Some crews operated their Ravens for ten hours straight in this fashion.

Ravens are typically flown by a crew of two, who have essentially the same jobs as the Predator team: one to fly the plane, the other to point the camera at the target. Unlike Predator crews, they are just ordinary soldiers who may have other specialities as well. Being a Raven operator is, like the drone itself, something you can just pick up and put down.

Unlike the Predator, which requires pilot's qualifications to fly, Raven operation can be learned in about three days. Operators learn about launching and recovery, and how to recognise and avoid various hazards such as trees, power lines, and occasional attacks by birds of prey – there are videos of Ravens being brought down by hawks and other predators. Keeping an awareness of helicopter flight paths is also important.

In Iraq, much of the catering was handled by contractors Kellogg, Root and Brown. This left some Army cooks with little to do, and at least one cook diversified and became his company's Raven expert. (10) That's a long way from the Top Gun "Best of the Best" elitism, even if Predator operators may be some way down the pilot hierarchy.

Users can also hone their skills on the ground control unit. Flying the Raven is said to be simple for anyone with experience of video games, although the industry tends to distance itself from any suggestion that operating a drone resembles a video game. The controller comes with a shrouded "viewing hood" to make the screen easier to see in bright sunlight – an echo of the black cloth that the TDR-1 operators covered themselves with in WWII.

The ground control unit can run training software, known as the Visualization and Mission Planning Integrated Rehearsal Environment or VAMPIRE (11). With VAMPIRE, an operator can practice flying virtual missions without needing to launch anything; it is like playing a handheld video game. An enhanced version can download sensor feeds from actual missions; this add-on is known as the Bidirectional Advanced Trainer (yes, that's VAMPIRE BAT).

Cruising at twenty-five miles per hour and three hundred feet, Raven is low enough and slow enough to catch important details. The key benchmark for drone video is telling the difference between an armed man and one who is carrying a shovel or other tool, and this is what Raven is designed to do. Being able to tell whether a position is occupied, or whether a group of men are working the fields or planting an IED, or where a mortar team is, makes all the difference to the commander on the ground. You can see the enemy – and they can't see you.

The video feed was originally recorded on a consumer eight-millimetre video recorder, a Sony Handycam, which allowed the user to freeze-frame or look back through the flight; it is now recorded digitally. The other piece of hardware is a ruggedized laptop, a Panasonic Toughbook computer. This provides a moving map display via Army software called FalconView. Raven operators, like their counterparts on the Predator, have one screen showing the drone's-eye view, and one showing the overall map view.

Apart from its ready accessibility to platoon commanders on the ground, the other key selling point of the Raven was its price. In 2012 a complete system with two ground control stations, three RQ-11B air vehicles, plus all the sensors, spares, and carry cases, can cost the US military $100 - $200,000. A single air vehicle on its own costs around $34,000. It is the sensor package, especially the thermal imaging, that pushes the price up.

To civilians that might seem like a lot of money for a radio-controlled model aircraft, but it needs to be put in context. In the conflict in Afghanistan, soldiers have on occasion used shoulder-launched Javelin anti-tank missiles costing $70,000 against individual insurgents behind cover. The mine-resistant MRAP armored trucks, hastily purchased to give protection against IEDs, cost about $600,000. The life-saving capability that the Raven offers is cheap at the price.

It's certainly a low-cost option compared to $14 million for a Reaper. The Reaper also costs about $4,000 an hour to fly, so one ten-hour flight costs as much as a Raven. The F-22 Raptor costs $50,000 an hour to fly, the F-35 over $30,000, making Reaper cheap by Air Force standards. This is why the Army's manual for the Raven lists one of its capabilities as being "attritable" – it does not matter if you lose one.

Bigger drones like Predator and Reaper need a full ground crew, not to mention that five-thousand foot paved runway and the facilities of an air base for maintenance and refuelling. All the Raven needs is a rucksack to carry it and somewhere to recharge the batteries; it comes with an adapter to charge from a Hummer or other vehicle. Maintenance and repair is carried out by the operator, thanks to the convenient modular design. Upgrading a Predator is a matter of sending it back to a major facility. Upgrading a Raven means posting a new component and having the operator plug it in. Again, the small drone looks more like consumer electronics than a traditional combat aircraft, even though it is carrying out an aircraft's tactical role.

The Raven Soars

The Raven proved highly popular with troops. One hundred seventy-nine sets, each with three Ravens, were quickly acquired by Special Operations Command in Afghanistan. In Iraq in 2005, when Special Forces teams from different services and different nations battled together against insurgents, the Raven was reportedly one of the most envied pieces of equipment around. The little drones were loaned out to various other units, and word soon spread.

US Army troops were among those who were impressed by their Special Forces colleagues' new gadget, and soon the Army was buying its own Ravens. By 2006, AeroVironment had shipped more than three thousand to units in Iraq and Afghanistan. The switch from small-scale production by hand to bulk manufacture may sound simple, but large-scale production turned out to be one of the most daunting challenges. When a system's fundamental requirements are robustness and reliability, quality control is paramount. Slip-ups at this stage could have destroyed the program, but AeroVironment successfully stepped up from small-scale R&D operation to full-scale drone factory.

Although it does not have a laser designator like the larger drones, which can mark targets for laser-guided missiles, the Raven can direct artillery or mortar fire. A typical example is shown on one YouTube video. Three insurgents are walking along a path, clearly unaware of the Raven flying just overhead; a second later they disappear in a burst of explosions. From their point of view the artillery fire came out of nowhere. The Raven achieves the scout's ideal of seeing the opposition without being seen.

The Raven can also carry out what is known as "battle damage assessment" after an air or artillery strike to confirm the effects of bombing.

The first version of the Raven was replaced by an upgraded model, the RQ-11B, in 2006. It had been enhanced by the addition of a laser illuminator to invisibly point out targets ("sparkle" them) to friendly forces. The illuminator is often used when working with allied helicopters. Once the Raven operator has a target, they can keep it illuminated until the helicopter can put a missile or gunfire on target. This might, for example, be a fleeing vehicle full of insurgents that cannot easily be distinguished from the other vehicles around it.

Improvements in geolocation – figuring out precisely where on the map the Raven's camera is pointing – make it possible to achieve "Category 1 target coordinates." This means the target is located to within twenty feet. New GPS-guided mortar bombs and artillery shells can be placed precisely with the first shot.

The RQ-11B Raven also has improved cameras and an endurance of up to ninety minutes on rechargeable lithium-ion batteries. There was also some help for patrols who have to go out looking for downed Ravens in the form of Falcon Tracker, a radio beacon to locate the fallen bird.

A steady stream of upgrades followed. The original infrared camera produced a rather murky image, but later generations are substantially better. Originally, the cameras were fixed, so the Raven had to be steered towards a target or orbit around it for the sideways-looking camera. Now there is a new gimballed mount called MANTIS that turns automatically to provide a stable image as Raven flies past a target. (12)

Previously, users had to choose between daylight and thermal imaging night cameras in the sensor bay, but MANTIS has both at the same time, with a resolution of five megapixels and a x4 optical zoom. It also adds an integrated laser illuminator.

This combination of daylight video camera, thermal imager, and laser illuminator makes MANTIS practically a miniature version of the payload carried by the Predator, although it does add somewhat to the price at around $18,000 a time.

The Raven's propeller blades sometimes underwent "field modification" in Afghanistan, and there has since been further work on the design. Options now include quiet propellers so the Raven can sneak in close without being heard, but also specially noisy propellers intended to attract the attention of people on the ground. These were useful for flushing insurgents out of hiding, provoking them into giving themselves away by firing, or simply driving them out of the area.

"It is a real deterrent for shooters wanting to get up on a roof and shoot at our guys," says one operator. "When we fly low, they can hear the plane; they know that we are there watching."(13)

And being watched by a Raven meant that American firepower might be brought to bear at any second, from artillery, helicopters, or other means.

"It does not take long for the enemy to associate the audio signature of the Raven with the lethal effects of USF [US Forces] action," noted one commander. Some operators even taped chemical light sticks to their Ravens on night missions to make them more obvious to the enemy. (14)

The same approach worked in Iraq. The Second Heavy Brigade Combat Team noted an inverse correlation between the number of Raven sorties they were flying and the number of attacks from insurgents. As Raven use increased, the attacks dropped from ten a month to two. The Raven's mission was described as "terrain denial," keeping the enemy away simply by flying around.(15)

Not everybody liked the Raven so well. A 2009 report by the Marine Corps found that the Raven was flimsy, and that the limited flight time meant that it could not be used for the same sort of "pattern of life" analysis as the Predator with its longer endurance. (16) Some Marines later worked around this limit by working the way Predators do – having a fresh drone take over each time the battery of the patrolling machine runs low. In one test, they maintained continuous surveillance for 24 hours with two Ravens.

The Marines also noted that Ravens had been lost, especially over urban areas, due to radio-frequency interference. Some of this had been from other drone communications using the same channel. Occasionally there was interference from Warlock radio-frequency jammers, which prevent bombs from being detonated by remote control.

The Raven went digital in 2009 with the introduction of the Digital Data Link (DDL). This allows sixteen Ravens to be flown in an area rather than just four, and goes some way to solving the interference problem. The digital version provides a securely encrypted video feed. By contrast, it was revealed in 2012 that more than half of the Predator and Reaper fleet were still sending unencrypted video. The digital link seamlessly connects the Raven to other systems, and allows Raven to act as a communications relay between different units on the ground or feed video to multiple users.

As with Predator and the portable ROVER display unit, which allows people besides the operator to see the drone's-eye view, there was a demand to share the Raven's imagery. AeroVironment now sells a pocket-sized device that turns any Android or iOS smartphone or tablet into a video terminal for the Raven signal. (17)

The Raven is starting to look like a thoroughly networked system. It would be possible, in principle, for the video feed from a Raven to be displayed in the White House Situation Room alongside images from its larger cousins.

Lockheed Martin has developed a system called VU-IT, which works with the Raven and many other platforms. For example, an Apache helicopter pilot with VU-IT can control a Raven, using it as a remote sensor to look into an area too hazardous to fly a helicopter over. So far, Ravens have belonged to the infantry, but there is increasing interest in flying them as "off-board sensors" so that helicopter pilots and others can also get a good view without exposing themselves to fire. (18)

It is not just the Raven's hardware that is being steadily improved. Software upgrades make life easier for the operator. For example, when the background itself is moving, it can be hard for the operator to spot small moving objects such as people on foot. Kestrel Moving Target Indicator software picks out and highlights moving objects in the field of view. Kestrel conveniently puts a marker rectangle around the moving objects – usually people and vehicles – making them obvious. (19)

Further software enhancements are likely to provide even greater capabilities, such as automatically following a vehicle or an individual, or identifying specific types of vehicles or equipment. A trained operator can spot an RPG launcher or a mortar base plate, but these are learned skills; in the future, object recognition software might takeover.

CCTV networks already have smart algorithms for tracking people and alerting the operator when something anomalous happens. For example, when monitoring car parks, they know that anyone who walks around for a few minutes has misplaced their vehicle or is looking to steal one. Similar algorithms could turn the Raven into a smart watcher that requires minimal supervision.

Raven is an outstanding example of a spiral development process. Rather than there being a Version 1 release followed by Version 2 and so on, there have been a host of small developments added incrementally according to requirements and the available technology. Some of these have been widely publicised, like the digital link; others, like a jam-proof GPS receiver are less well-known. But the improvements are there.

"Raven has had three major system level revisions over the years, and several hundred incremental enhancements," says Gitlin. "But the important thing was that the basic design was sound – and in particular, the modular system architecture that facilitates change."

The demand for Ravens continues to grow. Brigade Combat Teams in Afghanistan were initially issued with fifteen Raven systems, each with thee drones; commanders requested that the number of systems be increased to thirty-five to ensure there was one for every rifle platoon.

Raven has provided a radical new capability for ground troops. Gitlin is in no doubt what the most important aspect of its development has been:

"We can think of no greater achievement for Raven and its 'siblings' – Puma AE, Wasp AE – than the countless lives that have been saved by troops who were provided with this capability. These lives include the service members of American and the allied military forces that employ Raven as well as non-combatants who have been spared injury because of the situational awareness Raven, Puma AE and Wasp AE deliver."

He notes that the Raven has a wider significance in having opened the way for future development.

"Raven has forged a path for small unmanned aircraft systems…In the future, small UAS [drones] will deliver even greater capabilities in military and commercial markets."

An Aviary of Other Birds

The Raven has had rivals from the start. The RQ-14 Dragon Eye, adopted by the Marine Corps in 2004, was designed by the Office of Naval Research and, like the Raven, produced by AeroVironment. (20) It is a slightly heavier craft at six pounds, with a smaller wingspan of four feet and a propeller on each wing. The operator launches it with the aid of a rubber bungee cord, catapult-style. It has a shorter range and endurance than the Raven. The control system incorporates a pair of video goggles attached to a laptop computer.

The Marines initially favoured the Dragon Eye, which they considered to be more robust than the Raven, and ordered over a thousand of them. There were a number of upgrades, including improvements to the camera and communications. However, when the Raven RQ-11B was released, the Marines reassessed and switched to the lighter craft with better endurance.

Prioria's Maveric is favoured by some Special Forces. Maveric features an unusual flexible wing; this is rolled up for transport in a six-inch tube and unfurls itself instantly when it is pulled out. This makes Maveric faster to get into action than the Raven. Maveric has a quiet propeller and a silhouette that looks like a soaring bird. It takes an alert watcher to realise that it is actually a drone, making it popular for covert operations. (21)

As a lower-cost option, there is the Skate, made by Aurora Flight Sciences. This is just two feet across and weighs less than three pounds. It has an unusual design, a flat body made of rugged polyurethane foam and two propellers. It takes off vertically and then flips over from vertical to horizontal flight. Skate may be slower than other small UAS, with a top speed of fifty miles per hour, but it can hover and fly inside buildings. Take-off is also easy, and it can be launched from inside a building or a confined space, whereas Raven requires a short run when there is no wind. (22)

The Arrowlite is an Israeli drone that is starting to attract attention in the US military. Its vital statistics are similar to the Raven, but it can fly for almost three hours at a stretch (23). Meanwhile, aerospace heavyweight Lockheed Martin is also seeking to stake a claim in the small drone market with a new vehicle called Vector Hawk. (24)

There are also more shadowy small drones. Lockheed Martin's earlier Desert Hawk drone is favoured by the British Army, but also apparently by some US Special Forces. Pictures have been posted online of small drones camouflaged to look like birds that have come down over Pakistan. These appear to be modified Desert Hawks. There are also a couple of special variants, one equipped with an acoustic shot locator to pinpoint the location of a sniper from the sound of the gunfire. Another has a 3-D stereo camera system, developed specifically for locating and inspecting cave entrances in Afghanistan, which might otherwise be mistaken for shadows.

Perhaps the most notable rivals of the Raven are two craft made by AeroVironment – the RQ-20 Puma and the RQ-12 Wasp. (Those

RQ-designations shows that the drone is an approved military 'program of record' - a sign of official status rather than being a one-off purchase for a specific mission).

The Puma is effectively an upgraded version of the Pointer and offers longer endurance and a bigger payload than the Raven. The Wasp is the smallest of the lot, weighing less than a pound and with a wingspan of just over two feet, making it the most portable of the family and suitable for Special Forces and others who have to carry everything for days at a time.

There is a growing catalogue of manufacturers in the US and abroad making small drones for the military market. None of them has been as successful as AeroVironment to date, and users will always debate the finer points of superiority.

As the program of record, the "type classified" small drone of the US Army, the Raven is in a strong position. It would take something major to shift it from its place at the heart of the Pentagon's small drone fleet. For other nations, there are plenty of rival options available, and the technology is not so complex or expensive that it cannot be imitated with some effectiveness.

In 2011, Russian infantry units started to receive their first "Grusha" small drones, which appear to be similar to early-model Ravens in terms of performance. Russia now also fields the small "Granat-1". Other nations, including North Korea, also produce their own tactical drones. (25)

Ravens with Claws

Right from the start, Raven operators were frustrated that they could see the enemy but not do anything.

"You want to jump into the fight and you can't do it," one Raven operator told a TIME reporter after watching a firefight remotely. "It's bad when you're looking at it on the screen and you're looking at other soldiers getting shot at." (26)

Many made the obvious suggestion of fitting the Raven with an explosive charge, and in 2005 the British Daily Mail newspaper suggested that such a device was being used against the Taliban. (27) In fact the US Army's Lethal Miniature Aerial Munition System project (LMAMS) was started around this time, but took years to be put into action.

The first real signs came with mentions of a shadowy project known as Anubis. (28) Nobody would discuss it openly, but it was described in budget documents as a tactical micro air vehicle for engaging fleeing high-value targets: "The objective of this project was to develop, prototype, and test Non Line of Sight munitions with man-in-the-loop target ID with very low collateral damage as a proof of concept effort." By 2009 Anubis was described as achieving greater range and more precise lethality than any other squad weapon. Significantly, systems are also mentioned as being "deployed to Afghanistan awaiting an operational demonstration."

The follow-on to Anubis was the Switchblade, also made by AeroVironment. Rather than being hand-thrown, it is launched from a tube – the wings then flip out automatically, hence the Switchblade name. (29)

Switchblade resembles a disposable version of the Raven, having a similar size and speed. It has forward-looking daylight and infrared cameras, and a few extra features, such as a silent mode, turning it into a glider for additional stealth. It is also, of course, lethal: the operator locks on to a target and Switchblade pursues and strike the target, detonating high-explosive warhead as it impacts. The warhead is equivalent to a hand grenade and can destroy light vehicles such as pickup trucks or take out human targets within a radius of several metres. It can be directed through a specific window to hit the occupants of a single room. It even has a variable lethality setting, depending on whether the aim is to kill one individual without injuring others nearby or attack an entire group. This potential for fine-tuned targeting represents a radically different, surgical alternative to the Hellfire's destructive power.

Switchblade is arguably the most discriminating weapon ever made. It provides a close-up image of the target, and it has a "go-around" function so the operator can cancel the strike at the last moment. With other weapons – the rifle, the bayonet – the soldier may see the face of the enemy, but they will always be in potential danger themselves. It is hard to make a cool assessment of whether someone is a valid target if you might be shot the next second if you make the wrong call. The Switchblade operator can make their decision without the fight-or-flight rush of adrenaline.

The first few Switchblades deployed in Afghanistan in 2012 were a great success. For the first time, a soldier safely behind cover could seek out, locate, and neutralise a sniper from beyond line of sight. The whole engagement could be fought without ever being seen. A mortar on the other side of a hill could be silenced, and Switchblade could not only find out if insurgents were laying an IED but also stop them. Some four thousand have been deployed in Afghanistan.

The US Navy has also tested an extension of the Switchblade in which the drone is launched from a submerged submarine at periscope depth. It is released in an airtight capsule that releases the drone when it surfaces; the operator can then control it from inside the submarine as it carries out a strike against a surface target, either a ship or a target several miles inland. In theory, it could be used to carry out covert, precision strikes from the safety of international waters to targets anywhere close to the sea – including the many world cities with seaports.

The US Army plans to extend LMAMS and it may become a "program of record" in 2016 (29). This would make it a standard part of the Army's inventory. A 2014 proposal document requested pricing from manufacturers for up to 500 control units and 2,500 drones a year.

After good reviews from the field, AeroVironment's Switchblade is of course a strong contender for the LMAMS competition. But it is a small company, and now it's facing competition from industry giants including Raytheon, Textron Systems and Lockheed Martin.

Lockheed Martin's offering in this area is called Terminator. The original version had two sets of propellers, making it look like a miniature WWII bomber, and was launched vertically. Lockheed says that it is reliable and designed to work in varied weather conditions and lock on effectively. The system weighs less than twelve pounds. Interestingly, the airframe was produced on a 3-D printer, a trend we will look at in the next chapter. In 2015, a new version of Terminator was unveiled with a single pusher propeller; Lockheed Martin declines to comment on the new design.

Textron is big in the field of munitions. Textron's BattleHawk has a flight endurance of thirty minutes, so it can "loiter" or survey a wide area before finding a target. BattleHawk is said to be easy to use, with an Android-based user interface, a sign that open-source compatibility is becoming important. Textron's Cathy Loughman describes it as an evolutionary development: soldiers already have tactical UAS and use them to direct fires. BattleHawk puts everything in the same package.

Both systems have a top speed of over 100 mph and carry a warhead capable of engaging personnel and light vehicles. Terminator can act as a long-range precision grenade launcher, able to put a warhead into a foxhole or trench, or through a specific window from beyond line of sight.

Both manufacturers emphasise low collateral damage from high precision and the small warhead. In addition, they give a close-up image of the target before impact, and have the same "go-around" feature as Switchblade. (LMAMS future requirements may include 'non-lethal' warheads to disable targets without harming them, and there is even a proposal for a foul-smelling malodorant warhead to drive occupants out of buildings).

This type of weapon gives a game-changing new capability to the individual soldier, but the small drones are capable of far more. Both companies have been actively exploring the networking capabilities of the new weapons.

Other reconnaissance assets can send video and GPS data to the Terminator control unit. The operator can then launch Terminator and use the data to strike the target – and can transmit video and data to others. Similarly, Loughman says BattleHawk has been fully integrated into the US Army's Precision Fires Manager, a networked system using real-time data from multiple sensors to provide target data.

The US is not the only country developing this capability: the Israelis already have it. The Hero series of loitering drones made by UVision look a lot like LMAMS. They are launched from a tube, with flip-out wings and an electric propeller. The smallest, the man-portable Hero 30, weighs 3 Kilos (7 pounds) and has a flight time of thirty minutes. It is already in service with the Israeli military, and UVision is seeking other customers.

The implications of the lethal, portable drones for ground combat have not yet been worked out. We do not know what a firefight would look like if one side (or both) had a good supply of these drones. It is likely to be a transformative weapon, particularly when used in conjunction with the Raven or other small drones. Soldiers can see the enemy at a distance and attack them with incredible precision. If someone is shooting at you with machine guns or rockets, you can hide behind a wall or in a hole. There is no way of taking cover from LMAMS short of being inside a closed bunker. It can fly around to attack from any direction and can dive into trenches. This makes it unique, and the tried and tested tactics of previous decades will have to be revised. Lying flat and keeping your head down may simply make you an easier target.

In the future, firefights with rifles may become as rare as stabbing with bayonets. The real action may take place at longer ranges without any face-to-face contact at all.

The precision of LMAMS is also unique among infantry weapons. You have to fire a lot of unguided weapons to score a hit. For rifles and machine-guns, a figure of twenty thousand rounds or more is often quoted for every casualty. Artillery and mortars fire hundreds of rounds, tearing up vast areas of landscape without hitting a target. Guided weapons are different. From their first use in the 1960s, laser-guided weapons were hitting the target more than half the time. With such weapons, range is irrelevant: a target is as easy to hit from two miles away as from two hundred yards. A squad with twenty LMAMS is likely to score close to twenty hits – which may be on vehicles, firing positions, or individuals.

LMAMS will be a tremendous asset to US forces. But the technology will not be the sole preserve of any one country. The Israeli Hero will not be the only foreign rival. Like anyone else, US forces are likely to suffer heavy casualties if they come up against hand-launched lethal drones.

Moreover, Switchblade is the first generation, the primitive Version 1.0 of lethal mini-drones, a crude but useful new capability as Pointer was with reconnaissance drones. Future versions are likely to be much more effective, and more deadly. Given the pace of development and huge potential for small drones (as we will see in the following chapters) , it is impossible to predict what the battlefield of the future might look like or what role humans would play other than as warm targets.

Looking Forward

There was nothing inevitable about the success of the Raven. The Pointer did not set the world on fire, and without the development program that transformed it into the Raven, small drones might have been confined to a handful of Dragon Eyes and similar devices. But once Raven became established, it spread and evolved rapidly in its first decade in service. The current Raven looks a lot like the first models from 2003, but it is a very different bird.

Everything that the users have asked for – mission time, sensor resolution, ease of control – has progressively improved. The small drone is now in its own way as important as its larger cousins. A Raven at three hundred feet can send back imagery that is as useful as that from a Predator flying at ten thousand. This is not to say that a Raven can do everything a Predator can -- at least not yet -- and the two are certainly not in competition at this point. But when it comes to providing answers to key questions – Is this our target? Are these insurgents or farmers? Where is the mortar firing from? – the Raven is already invaluable, and getting steadily better.

Being under direct control, and hence able to beam back imagery right now and not in an hour's time, is a tremendous advantage. Being able to share that information with other users on the ground magnifies that advantage.

As we shall see, the upgrade process is likely to continue at an accelerated pace. This is in sharp contrast to larger drones. Upgrading a Predator is an expensive business. You will need several million dollars of aircraft to test it on, and integration with existing systems is complex. A new piece of hardware is likely to be expensive, and because your total market is only a few dozen aircraft, the R&D costs will be high. But the Raven is modular; you can unplug one unit and plug in another. If you develop a new piece of kit that can, for example, measure wind speed, then some of the many different users are likely to be interested, and you have a market. You can mass-produce and make things cheaply.

The huge and growing existing user base means that there will be a ready market for small drone developments in a way that could never happen with their larger cousins, as we will see in chapter 6. Before then, in chapter 5, we will look at the solution to the biggest limitation of the Raven and its predecessors and rivals: flight time.

But before we leave this chapter, we need to continue another revolution in small drones that has happened largely outside the military sphere.

A Parallel Universe: The Evolution of Quadrotors

Fixed-wing drones like the Raven have a competition in the form of electric-powered multirotor helicopters. These are very different from earlier helicopter drones.

The US Army spent considerable effort on a small helicopter drone. This was the RQ-16 T-Hawk (for "Tarantula Hawk," a type of wasp) made by Honeywell. Known as the "flying beer keg," this is a twenty-pound cylinder, two feet long, wrapped around a ducted fan powered by a gasoline engine. (30)

The ducted fan is an enclosed propeller; being inside a casing means there are no stray rotor tips to catch on vegetation or buildings. Developed under a $70 million contract, the RQ-16 was originally part of a fantastically ambitious program called the Future Combat System, which envisaged a whole menagerie of robots on the ground and in the air working alongside tanks and foot soldiers. The FCS planned to combine all the manned and unmanned vehicles in one unified network that would work and fight together as one. However, while the vision may have been sound, the technology was not, and it was terminated in the 2000s when the military version of the dot-com bubble finally burst.

The T-Hawk, the sole survivor of the cull, went into low-rate production, and was dispatched to Iraq for testing by the Army. A number of modifications followed, including a fuel-injected engine for easier take-off, better flight control software, and a sensor turret rather than fixed cameras. As with the Raven, a digital data link version followed in 2010.

The T-Hawk proved useful for finding IEDs, largely thanks to its ability to hover in place, and is still used in this role by the British Army. However, the US Army officially canceled it in February 2011 in favour of fixed-wing drones. The main reason seems to have been noise. The Raven is audible when it flies overhead if the operator wants it to be heard; the T-Hawk was loud and could be heard from hundreds of yards away. Any enemy could get under cover long before it was close enough to see them.

In addition, unlike the Raven, the T-Hawk is not readily portable; at a basic twenty pounds, it cannot be tossed into a rucksack. Refuelling was more difficult than plugging into a charger. The flight time of forty minutes cannot easily be extended. Perhaps more important, the cost, at something over $500,000 per vehicle – about fifteen times as much as a Raven – means it is not expendable as the Raven is.

However, there is now a smaller, quieter, cheaper alternative. Recent years have seen tens of thousands of miniature rotorcraft take to the skies, which can be piloted from a smartphone. Unlike normal helicopters with one horizontal rotor blade for lift and one vertical rotor for stability, they have several sets of rotor blades that all provide lift.

A helicopter is necessarily a complex, delicate, and expensive piece of equipment. This is because steering involves changing the angle or pitch of the rotor blades, which needs an elaborate mechanical arrangement. The quadrotor has four sets of blades, and steers and maintains stability simply by speeding up or slowing down different rotors. Without modern electronics, it would be impossible; with them it is easy.

Modern multirotors date back to the late 1980s with the Gyrosaucer toy produced by Japanese company Keyence. However, modern developments tend to be traced back to US engineer Mike Dammarm, who developed his first battery-powered quadcopter in the early 90s. This was marketed by Spectrolutions Inc. as the Roswell Flyer in 1999 and later adapted into the Draganflyer, a range which is still going. (31)

Combined with the latest in small video cameras, multicopters could send back the sort of pictures previously only possible with a helicopter. Multicopters multiplied, and the big breakthrough came in 2010 when Parrot produced their first AR.Drone. This was hailed as a fantastic toy: a helicopter sending back video via Wi-Fi which you control with an iPhone. The AR.Drone was a bestseller, and the world woke up to the potential of multicopters.

Larger professional versions are already in widespread use by police forces and film crews, capturing dramatic footage in *SPECTRE, Avengers: Age of Ultron*, and in numerous TV documentaries. Scientists use them to track wildlife or carry out land surveys. Engineers inspect power lines, bridges, and power station chimneys via drone. Their biggest limitation is the legal situation in the US; the FAA has so far refused to licence them for commercial activity, with few exceptions.

One of the most significant companies in the business is Aeryon Labs which produces a series of drones including the popular Aeryon Scout. This weighs three pounds without payload and has a maximum speed of thirty miles an hour, carrying daylight or thermal imaging cameras. Like the Raven, it has an easy user interface with a game-type controller. It can be set to hover in place and "stare" at an object of interest at the touch of a button. In April 2014 the company announced an order from an unnamed US military customer.

"Our VTOL sUAS (Vertical take-off and landing small unmanned air system) platforms are field tested and battle proven" said Kevin J. Kane, President and CEO of Datron, a claim repeated on the company website, indicating that there had been previous, unannounced use by the same customer. (32)

The Aeryon Scout is another modular system, with a camera pack that can be easily upgraded. The latest version, the HDZoom30, features a powerful zoom lens for both civil and military operators. For example, it can read the serial number on a power line insulator at a hundred feet; it can also read car licence plates and recognise faces at a thousand feet. (33)

Like the Raven, the Acryon Scout is backpack portable and can be launched in minutes; also like the Raven, it is a tenth of the cost of the T-Hawk and very much quieter.

AeroVironment would prefer the Army to have something even more like the Raven: the Shrike, their own new quadrotor, which uses many of the same components and the same controller as the Raven, Puma, and Wasp. It was introduced in 2011, but has so far failed to attract significant orders.

Multirotors have a couple of major advantages over fixed-wings. For one thing, they can operate indoors, going through buildings, tunnels, and bunker complexes. And they give a more stable view than a moving fixed-wing.

The future may see a mixture of fixed-wing, Raven-type craft used for long-range missions and quadrotors for close-in work and situations where being able to hover motionless gives a big advantage. But it may be possible to combine the two in a single platform that gives the best of both.

Lockheed Martin's Vector Hawk can be configured as a fixed wing or as a multirotor depending on the mission. It can even be turned into a tilt-rotor combining vertical take-off with fast forward flight like the Skate. Like Raven, Vector Hawk weighs four pounds, and the makers claim that it has the "best-in-class payload capacity, speed and endurance." (34)

Others claim they can get better performance with novel designs. British company VTOL Technologies, based at the University of Reading, have an airframe that produces high lift at low speed, and low drag at high speed. (35) This design, the Flying Wing, has two pairs of ducted fans which can rotate to any angle, so they can propel the drone forwards in level flight or provide vertical lift. What distinguished the Flying Wing from other tilting approaches is that the wing can be pointed into the wind to provide most of the lift needed for hover – gulls and other soaring birds use this technique to hover against the wind. This can extend its endurance to several hours.

The Flying Wing's innovative airframe is supplied with a precise navigation system, automated collision-avoidance, and a beyond-visual-range control system. The prototype Flying Wing has a three-foot wingspan and is currently undergoing flight-testing. The makers suggest that it would be suitable for delivering high-value packages, while the high precision and efficient hover make it good for jobs like infrastructure inspection – getting close-up pictures of power-line equipment or industrial chimneys. There is also a military model that combines Raven-like speed and endurance with the precision and vertical landing of a quadrotor.

Meanwhile, a start-up called Krossblade based in Tempe, Arizona, is taking a more radical approach with their SkyProwler. (36) This is a transformer drone: in forward flight it looks like an airplane, but four rotor blades flick out of the body for vertical take-off and landings. This combination of two types of rotor make it highly efficient at both hover and forward flight, with the only penalty being the additional weight of the transformer mechanism and rotor blades. CEO Dan Lubrich estimates that the mechanism only adds about 10-12% to the overall weight, and provides major benefits.

In spite of its small size, the SkyProwler can cruise at over sixty miles an hour with a top speed of over eighty. It has an endurance of forty minutes with a one-pound payload. Krossblade was taking pre-orders in 2015 with the aim of delivering SkyProwlers by March 2016 at under $3,000 each.

Whatever the exact form they take, small drones will become increasingly capable as successful innovations are taken up by the rest of the industry. They may be some sort of tilting design that can switch between vertical and horizontal flight, or they may be able to transform or reconfigure themselves in some way. The different technologies will battle it out in the marketplace, and a winner may emerge. The winner may not be best design, but the one with the most backing. Whatever happens, the future small drones are likely to be able to hover, take off vertically, and explore confined spaces like building interiors just as the Black Hornet can now.

REFERENCES

1) "…nine thousand of them are small, hand-launched craft made by AeroVironment of California."

Hambling, D. Raven shows decade of advance for small UAS.

Aviation Week. (http://aviationweek.com/awin/raven-shows-decade-advances-small-uas)

2) "Dr. Paul MacCready Biography"

Dr. Paul MacCready Biography. (2007).

https://www.avinc.com/about/dr_maccready/bio/

3) "RQ-11B Raven" product page, AeroVironment

UAS: RQ-11B Raven. (n.d.).
(http://www.avinc.com/uas/small_uas/raven)

4) Pointer Evolution: "Special Ops Using 'Pointer' UAV for Vital Recon, Force Protection"

Special Ops Using "Pointer" UAV for vital recon. (2013, July 11). Retrieved from
(http://www.uadrones.net/military/research/2003/0711.htm)

5) Pathfinder Raven

Mawn, A., & Tokumaru, P. (2004). The Pathfinder Raven small unmanned aerial

Vehicle. Retrieved from (http://1.usa.gov/1QWyijH)

6) "The Little Plane That Could"

Martini, Noma. "The Little Plane That Could." *Office DoD Science Blog.*

Armed with science. (2010, November 9). The little plane that could [Blog post]. (http://science.dodlive.mil/2010/11/09/the-little-plane-that-could)

7) "I now have something as a company commander to give me three-dimensional situational awareness."

Army Magazine (http://www3.ausa.org/webint/DeptArmyMagazine.nsf/byid/CCRN-6CCSFL)

8) "Like a cartoon"

McClary, Ruth. "Ravens support U.S., Iraqi forces." Defense Video & Imagery Distribution System.

McClary, R. (2009, August 17). Ravens support U.S., Iraqi forces. https://www.dvidshub.net/news/37568/ravens-support-us-iraqi-forces#.U25Wldq9KSM (http://bit.ly/1Ns02pF)

9) "The perception is that this [catching] will minimize damage" – Steve Gitlin personal communication October 2014

10) "Army cooks with little to do, and at least one took the opportunity to diversify as his company's Raven expert."

Blake, K. (2005, April 15). Ravens. (http://bit.ly/1NJC9QI)

11) Vampire

Vampire. (2004). (http://www.avinc.com/uas/vampire-flight-simulator)

12) MANTIS

Mantis - Revolutionary vision awareness. (n.d.). (http://www.avinc.com/uas/gimbaled_sensors)

13) "It is a real deterrent for shooters wanting to get up on a roof and shoot at our guys."

Blake, K. (2005, April 15). Ravens. Retrieved from (http://bit.ly/1NJC9QI)

14) "It does not take long for the enemy to associate the audio signature of the Raven with the lethal effects."

Martini, Noma. "The Little Plane That Could." *Office DoD Science Blog*. (http://science.dodlive.mil/2010/11/09/the-little-plane-that-could/)

Armed with science. (2010, November 9). The little plane that could

[Blog post]. (http://science.dodlive.mil/2010/11/09/the-little-plane-that-could/)

15) "Terrain Denial"

Harris, N. (2010, November 9). A Raven to make Poe proud. (http://bit.ly/1Ns29d7)

16) 2009 Marine report finds Raven flimsy

Ackerman, S. (2011, May 12). Handheld Spy Drone Too Wimpy for Iraq's Marines. (http://www.wired.com/2011/05/handheld-spy-drone)

17) Digital Data Link

Digital data link. (n.d.). (http://www.avinc.com/uas/ddl/)

18) VU-IT

VU-IT. (n.d.). (http://www.lockheedmartin.com/us/products/vuit.html)

19) Kestrel moving target indicator

Kestrel - Land MTI for small unmanned aircraft systems. (n.d.). (http://www.avinc.com/uas/kestrel)

20) Dragon Eye

US Army close to decision in $400m small UAV contest. (2015, September 27). (http://www.flightglobal.com/news/articles/us-army-close-to-decision-in-400m-small-uav-contest-201726)

21) Prioria Maveric

Prioria Maveric. (n.d.). (http://www.prioria.com/maveric)

22) Skate

Aurora Flight Sciences official product page (http://www.aurora.aero/Products/Skate.aspx)

23) Arrowlite

Arrowlite. (n.d.). Retrieved from (http://starkaerospace.com/arrowlite)

24) Desert Hawk

Desert Hawk III. (n.d.). Retrieved from
http://www.lockheedmartin.com/us/products/desert-hawk.html

25) Russian "Grusha" drone

Drones take off in Russia's armed forces. (2011, December 19). (http://sputniknews.com/voiceofrussia/2011/12/19/62504769)

26) Thomson, Mark. "The New U.S. 'Smalls' Air Force Over Afghanistan." *TIME*.

Thompson, M. (2011, June 26). The new U.S. "Smalls" Air Force over Afghanistan. (http://nation.time.com/2011/06/26/the-new-u-s-smalls-air-force-over-afghanistan)

27) WASP with explosive as Talibanator

Hambling, D. (2007, January 23). Military Builds Robotic Insects. (http://archive.wired.com/science/discoveries/news/2007/01/72543)

28) Project Anubis

Hambling, D. (2010, January 5). Air force completes killer micro-drone project. (http://www.wired.com/2010/01/killer-micro-drone)

29) Switchblade

Switchblade. (n.d.). (https://www.avinc.com/uas/adc/switchblade)

30) LMAMS

Small UAS Expand Air Support Options for US Army. (n.d.). (http://aviationweek.com/awin/small-uas-expand-air-support-options-us-army)

31) RQ-16 Taratula Hawk

Route clearance goes high-tech. (2010, December 1). Retrieved from (http://bit.ly/1TAL0mT)

32) Quadrotor evolution

Tracing the origins of the multirotor drone, for business and pleasure. (2014, November 2). Retrieved from (http://www.engadget.com/2014/11/02/tracing-the-origins-of-the-multirotor-drone)

33) "Our VTOL sUAS [Vertical take-off and landing small unmanned air system] platforms are field tested and battle proven."

Datron world communications. (n.d.). Retrieved from (http://www.dtwc.com/products/unmanned-solutions/vtol-suas)

34) Aeryon Scout

Aeryon Scout – Aeryon Labs Inc. (n.d.). https://aeryon.com/aeryon-scout

35) Lockheed Martin Vector Hawk

(http://lmt.co/1Q39o1y)

Lockheed Martin introduces latest addition to small unmanned aircraft system family. (n.d.). (http://lmt.co/1Q39o1y)

36) VTOL technologies Flying Wing

The flying wing. (n.d.). (http://www.vtol-technologies.com/flying-wing.html)

37) Krossblade Sky Prowler

Krossblade Sky Prowler. (n.d.). Retrieved October 23, 2015, from (http://www.krossblade.com)

CHAPTER FOUR

A SHARP FALL IN THE COST OF VICTORY

> *"There is no victory at bargain basement prices."*
>
> - Dwight D Eisenhower

A hand-launched Raven drone costs tens of thousands of dollars. It may look like a radio-controlled drone toys costing a few hundred but is far more capable and robust. A new class of commercial and "prosumer"drones has started appearing between the two drones which cost a few thousand dollars and are rapidly eroding the distinction between military grade and consumer grade.

Many such drones are sold for commercial and industrial use. The biggest sector is agriculture, where drones can spot areas that need water, fertilizer or pesticide more easily than a ground survey. Others verge on the military. The Lehmann 960, for example, is a fixed-wing drone with a three-foot wingspan for surveillance use by "security professionals." Like the Raven, it gives a real-time video feed from video and thermal cameras, but it is yours for just $7,000. (1) Professional photographers and movie makers use DJI quadrotor drones costing less than half as much and get spectacular results.

Military-grade drones could soon be very cheap indeed, making them different from other military aircraft that follow a sharp upwards curve of price. Rather, small drones are progressing along the path of smartphones and other consumer electronics, getting more powerful and more capable – and cheaper.

Military drone makers emphasize quality, and are keen to distinguish between their products and those flown by hobbyists. But we will see how a team at MITRE Corp built a military-grade drone called Razor from commercial components for just $2,000, boasting "90% of the capability at 10% of the cost."

This is revolutionary for military aircraft procurement. For decades, aircraft inevitably have become more expensive as they get more capable. To understand this situation, you have to take a step back and look how aircraft have developed since the Second World War.

Higher, Ever Higher: Theme with Variations

In 1984, Norman Augustine, former Under Secretary of the Army, and CEO of aerospace company Martin Marietta, published a set of "laws" about military procurement. (2) These were humorously cynical aphorisms derived from many years in the defense business. Some are relevant to any big organization, like Law 26 – "If a sufficient number of management layers are superimposed on each other, it can be assured that disaster is not left to chance."

His most celebrated pearl of wisdom is Augustine's Law 16, which says that the cost of each new generation of military aircraft rises exponentially. Augustine predicted that this would reach its peak in the year 2054, at which time:

"The entire defense budget will purchase just one aircraft. This aircraft will have to be shared by the Air Force and Navy 3-1/2 days each per week except for leap year, when it will be made available to the Marines for the extra day "

Although intended facetiously, Augustine's Law 16 has been remarkably accurate. The North American P-51 Mustang was one of the most important US fighters of WWII. Over fifteen thousand were built, at a cost of around $50,000 each in 1945 dollars ($655,000 in 2014). It was succeeded in the 1950s by the jet-powered F-100 Super Sabre at a cost of $700,000 – ($6 million in 2014), ten times as expensive in real terms. The McDonnell-Douglas F-4 Phantom, which first flew in 1960, broke the million-dollar barrier, costing $2.4 million apiece in 1965 ($18 million in 2014), tripling the cost of its predecessor. Even allowing for inflation, the upwards curve is steep.

By the 1980s, when Augustine was writing his laws, the escalating costs were obvious to everyone. The USAF's new F-15 Eagle, also from McDonnell Douglas, was set to replace the F-4. The Eagle was a superb aircraft, but it had reached a new high, costing in excess of $20 million ($45 million in 2014), almost tripling again the cost of its predecessor.

Because the F-15 was so expensive, only a thousand were ordered, when there had been five thousand F-4s. The F-15 was designed to excel at kills "beyond visual range," swatting enemy aircraft with hefty radar-guided missiles before they knew it was there. Smaller aircraft that could not carry this type of big, long-range missile would be completely outgunned; hence the F-15 had to be a substantial aircraft with a corresponding price.

Unfortunately, when the enemy gets to close range, air combat follows a different set of rules.

Extensive flying exercises found that the big twin-engine F-15 was only slightly superior to the small, cheap fighters fielded by the Russians in a dogfight. If it came to a war, the small band of F-15s would be overwhelmed by swarms of Russian MiGs. Certainly, the F-15s would be able to knock out plenty of the Russians at long range, but when the survivors closed with them, the contest would be bloody and one-sided.

The Air Force decided to go for a "high-low" mix, supplementing the elite F-15s with a large number of cheaper aircraft known as lightweight fighters. The aircraft selected for the lightweight fighter role was the single-engine F-16 Fighting Falcon, two-thirds the size of the F-15. It was to be the embodiment of a concept by fighter guru John Boyd for an austere aircraft with extreme agility that could beat anything in a dogfight. Being less complex, it would be so cheap it could be acquired in vast numbers. The F-15 with its powerful radar was the champion at long-range combat; the agile F-16 was to be the champion in the "furball" of dogfighting.

During the development process, the purity of the F-16 was slowly corrupted. It became heavier, less agile, and more expensive as more and more capabilities were added. Not everyone shared Boyd's vision. Rather than being a dedicated dogfighter, the final product was a multi-role aircraft. It was capable of accurate bombing as well as air-to-air combat; it was so successful as a precision bomber, the development of laser-guided weapons was held up as it was doubted they were necessary. The F-16 became powerful enough to carry large, radar-guided missiles for the interceptor role like those on the F-15. The F-16 was a good plane, but not what had been originally envisaged. Some European partners in the Falcon consortium were not impressed.

"Whatever happened to the lightweight, low-cost fighter we signed up to buy?" demanded the Belgian representative at one meeting. (3)

At $15 million in 1998 dollars ($22 million in 2014), the F-16 was cheaper than the F-15, but more expensive than anything in the previous generation, including the big F-4. Cutting costs was harder than anyone except Augustine had realized.

The US Navy went through a parallel experience. They also replaced the F-4 Phantom, and chose the F-14 Tomcat, a $38 million (1998 dollars, $55 million in 2014) carrier-based fighter. Like the F-15 it had a big radar and impressive long-range capabilities.

Again the F-14 was too pricey to acquire in large quantities, and the Navy took up the idea of bolstering numbers with a smaller, cheaper aircraft. They chose the F-18 Hornet, originally a failed competitor in the Air Force's lightweight fighter competition. The F-18s costs grew from a planned $5 million to around $29 million (2003 = $37 million in 2014). Building cheap had not worked out, and Augustine's law remained stubbornly unbroken.

Aircraft prices rose inexorably, but US lawmakers seem content to keep increasing spending. Numbers have fallen, but Augustine's prediction a single-plane Air Force will not quite come true because the defense establishment remains willing to pay higher prices. Shrewd defense contractors have located their plants across many different states to ensure continued political support as important local employers.

Still, efforts were being made to stem the tide of rising costs. When the Advanced Tactical Fighter was announced to supersede the F-15 Eagle, a key point was that it should cost less than the aircraft it replaced.

The Advanced Tactical Fighter program office set a goal of $35 million per plane in FY 1985 dollars ($66 million in 2014), less than the latest models of the F-15. New technology meant that the F-22

Raptor, as it became, would be the first generation to be cheaper than its predecessor. Augustine's Law was going to be beaten once and for all. (4)

Needless to say, it did not work out like that. The Raptor program eventually swallowed over $60 billion of the taxypayer's money, far more than the original estimates. This was in spite of the fact that the number to be purchased dropped from 800 to 400 and finally less than 200 aircraft. The "flyaway cost" for each aircraft is quoted at $150 million in 2008 dollars ($164 million in 2014). However, this does not include the sunk costs of research and development. If you divide the program cost by the number of aircraft delivered, you get a staggering $300 million per plane. Or about six times as much in real terms as the previous generation.

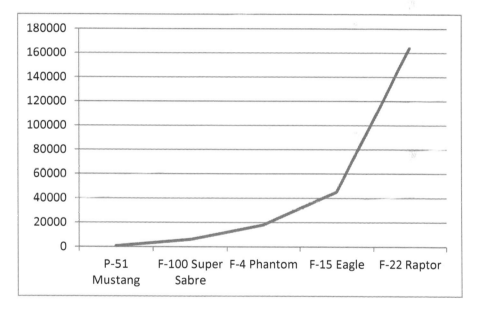

Since then, any arguments about the cost of the Raptor have been drowned out by arguments over the F-35 Lightning II. This is the replacement for the F-16 Lightweight Fighter. Like its predecessor, the F-35 is a single-engine aircraft that was originally meant to be a relatively low cost companion to its bigger partner. Low cost is of course relative; in 2001 Lockheed Martin projected a cost of $50 million a plane in 2001 dollars ($67 million in 2014).

The reader will not be surprised to learn that the cost, weight, and complexity of the F-35 rose inexorably during the development process. This has given rise to a whole ecosystem of lobbyists, enthusiasts, and apologists, as well as critics, who have mountains of cost data to juggle with. In 2015 there are still furious and elaborately detailed arguments about the actual cost.

As of 2014, eight years into production, the Air Force's version of the F-35 cost around $190 million in 2014 dollars, more than any other fighter except the bigger F-22. The UK is paying about £180m / $280m each for its first batch of fourteen. The total 'cost of ownership,' including maintaining and supporting a F-35 over its lifetime, is currently estimated at over $300m per plane.

Makers Lockheed Martin claim costs will go down as aircraft production ramps up and the aircraft matures over the next decades. Critics suggest that the price is as likely to rise. They also doubt whether the F-35 will be as capable as first envisioned.

In any case, the Pentagon's best guess is that the overall cost of the F-35 program will be in the region of a trillion dollars. At this point though, there is no alternative. It has taken too long and cost too much to get this far. The first development contract was signed in 1996, and the first F-35 flew in 2006, finally getting into service in late 2015.

With a development cycle of almost twenty years, starting over again and trying again is not an option. The F-35 program is too big to fail and too important for national defense to be cancelled.

It is easy to be too critical of an individual program. Many of the same criticisms levelled at the F-35 could be made against any fighter aircraft from the last forty years, and while the F-35 may not be especially good, it is certainly on-trend as far as the cost growth is concerned. It is not an anomaly, and it would be unrealistic to expect that if a different aircraft had been chosen the cost would have been any less.

The F-35 is simply another example of Augustine's Law.

It is one thing to observe that aircraft costs keep rising. It is another matter to explain it, and that would require another book. One simple factor is that the price of an aircraft is related to its weight. Weight adds capability; higher speed demands bigger engines, longer range requires more fuel, and both mandate a bigger aircraft, as does increased bombload. Each generation is bigger.

The 9,000 lb Mustang of WWII has bulked up into a 64,000 lb Raptor.

Another driver of cost is the quest for excellence at any price, because nothing else is worth having. As a pilot in *Top Gun* puts it: "Remember, boys, no points for second place."

The Soviets preferred investing in several cheap aircraft to one expensive one, flown by pilots with less flying time, and counted on winning by weight of numbers. But the US Air Force does not plan to sacrifice large numbers of pilots by giving them inferior aircraft. They want an elite force equipped with high-tech capable of destroying more numerous opposition.

The F-22 has unique superiority in three areas. (5) It is stealthy, making it impossible to spot at long range on radar, so a Raptor can pick off opponents (or avoid them) before it can be detected. Vectored thrust – being able to change the direction of the jet exhaust – makes the Raptor hypermaneuverable and able to outmatch any opposition in a dogfight. And sustained supercruise – flying supersonic for a protracted period, rather than in bursts – means that the Raptors can choose when and where to fight, breaking off from an engagement when it suits them. Its missiles are stored away in internal weapons bays, which maintains stealth and keeps drag down for greater speed and agility.

In combat exercises, four F-22s with experienced pilots can take on twelve opponents and win. Raptor pilots claim they can take on so many opposing planes that they run out of missiles and have to go in with their cannons to take out more before exiting the fight safely. Odds of three, four, or five to one when fighting inferior aircraft do not worry them, and anyone up against Raptors in an inferior aircraft is likely to be in trouble. In these terms, the Raptor is an excellent investment in guaranteeing air supremacy in years to come.

However, the need for excellence really pushes the price up. It is much the same as in Formula 1 race cars, where designers go to extremes to squeeze out every last drop of performance. The Pareto Principle, which is often quoted in business circles, is also known as the 80:20 principle and manifests itself in many ways. 20% of your customers account for 80% of the business, 20% of the bugs generate 80% of the problems and so on. A saying in defense circles has it that adding an extra 20% of capability accounts for 80% of the cost. Augustine's Law 15 is more pessimistic, saying that "The last

10% of performance generates one third of the cost and two-thirds of the problems."

As in racing, better performance is worth the price. A winning car, or a fighter that can shoot down the enemy fighter, is good value for money. Analysts call them "exquisite weapons": finely engineered at a fantastic price.

This drive for excellence leads to expensive items with short production runs (we will see later how this affects electronics). It may also lead to abuses by contactors. While many items – including the RQ-11B Ravens – are bought on fixed price contracts, many aircraft are 'cost plus' deals. The contractor is guaranteed to make a profit, however much it costs them to manufacture the article ordered.

In the 1980s the Packard Commission famously discovered the Pentagon was paying $435 for a hammer and $7,000 for a coffee pot. While people may not know how much a specially-made precisely-engineered titanium rivet should cost, these were obviously insanely high prices for common household items.

Such high prices are not entirely a matter of contactors seeking to rip off Uncle Sam. Sometimes they are a consequence of the procurement process. The Packard Commission found that the ashtrays on the E-2C Hawkeye aircraft cost $659 each ($1,446 in 2014). Younger readers may be surprised that smoking was allowed on military aircraft in the 80s. (7)

The cost was supposedly because of the government specifications: each ashtray was made of eleven parts and took thirteen man-hours to construct. The specification ensured the ashtrays would survive the jolt of landing on an aircraft carrier. The tiny production run meant that the unit cost was high...although perhaps not that high. Negotiation brought the unit price down to a more reasonable $49 ($107) – still pricey compared to an off-the-shelf alternative, or an old mayonnaise jar as suggested by the then Secretary of Defense, Caspar Weinberger.

The spiraling cost of military aircraft is not a uniquely American problem rooted in the nation's famed "military-industrial complex" as some might suggest. Europe's Multi Role Combat Aircraft program experienced similar mission creep and cost growth. The cost rose from a planned $5 million in 1970 to an actual $20 million by 1985 when the Tornado was delivered. (The Spitfires that fought in the battle of Britain cost about £5,000/ $20,000 in 1940 -- £84,000/ $126,000 at today's rates). Numerous books and TV documentaries criticized the program, highlighting it as an example of European bureaucracy and inefficiency. While the multinational setup did not help, the overall outcome was remarkably similar to the US experience.

The Tornado's successor, the Typhoon, was a conscious effort to make a low-cost aircraft as an alternative to buying an expensive US-made plane.

In 1988, the cost to the UK for 232 Typhoons was estimated at around $35 million each. By 1997 this had almost doubled, and by 2003 it had tripled to $105 million per aircraft. It might well have been cheaper to buy American. But as in the US, politics counts as well as practicality. The program maintained thousands of jobs in Europe and created a possible export earner.

Undoubtedly, producing an outstanding aircraft – or even an adequate one – is an expensive business. A side effect is that because they are expensive to replace, military aircraft cannot be traded in every few years for a better model like cars. They have to give decades of service, and inevitably that too drives the price up. The average USAF aircraft is now more than 23 years old. The last F-16 was delivered in 2005, and the last F-15C dates from 1985. Some of the bombers are real old-timers: by 2012 the average B-52 bomber was over fifty years old.

There are numerous upgrades and enhancements, but these tend to come at intervals of several years, and sometimes longer. The upgrades themselves run to millions of dollars per plane. The initial procurement cost of an aircraft may be dwarfed by the cost of the subsequent upgrades, especially when these include elaborate electronics.

The F-15 radar, for example, has been through four versions since the 70s. The latest version, the AN/APG-63(V) 3 was developed in 2006. This is solid-state, electronically scanning radar with no moving parts and advanced processing to identify and track more targets at greater ranges than previous interations. Improved discrimination means it is better able to see stealthy targets and to cut through clutter and radar jamming. The radar costs around $8 million a time, and at nine years old, it is in military terms a modern piece of equipment. Especially compared with other parts of the same plane.

The cost of the latest update to the twenty B-2 Spirit stealth bombers is a shade under $10 billion. The only alternative to paying would be to lose a key component of the US strategic bomber fleet.

The Predator and Reaper fit right into this picture, with the first-generation Predator being replaced by something that was faster and more capable as well as heavier and more expensive.

Given the effect of Augustine's Law, you might expect the next generation of small drones to be pricier and more capable than the current Raven. The standard pattern would have the drone getting bigger until it was no longer man-portable, growing over generations to resemble a small stealth fighter costing a million dollars or more. Instead of ten thousand Ravens, the Pentagon would have a few hundred mini super-drones, just as they have moved from fifteen thousand Mustangs to a few hundred Raptors. Some would claim that this was a great improvement, with perhaps no more than a little nostalgia for the old days.

However, there are compelling reasons for believing that drone development will follow an equally well established law from a different field.

Moore's Law Strikes Back

Away from the military, procurement is not ruled by Augustine's Law. In computing and electronics, purchasers and planners work to Moore's Law, which says that the price for a given capability goes exponentially down. This law was formulated in 1965 by Gordon Moore, a cofounder of the company NMI which was to become Intel, the world's most successful maker of silicon chips.

Moore's Law states that the number of transistors on a chip doubles every two years (a variant says eighteen months). This was based on Moore's observation of the industry, with each generation of chips being made with more advanced technology on a smaller scale. Given that the cost of the chip remains about the same, computers get steadily more powerful or cheaper as the cost of processing power goes down.

Moore's Law has proven accurate, to the point where it has become self-fulfilling: developers use it as a guide to planning the next generation of chips. Processing speed and memory increase, and price per unit of computing power drops. Similar laws also apply to related fields, such as mass storage or the number of pixels per dollar in digital imaging.

Moore also posited a corollary to his law – that the cost of each generation of production machinery would rise at a similar rate. The cost of chip production facilities or "fabs" has increased in line with this law. This does not affect the consumer because chips are produced by the tens of millions and the fab cost is spread out. When your entire production run comprises two hundred of a particular avionics system for an F-22 Raptor, there are no economies of scale and the customer has to carry the entire cost. This makes specialized electronics are so pricey, and is a contributing factor to the high cost of military aircraft.

In the 2010s some devices have reached a plateau in processing power. Every laptop, tablet, and smartphone is powerful enough for virtually everything the user wants and there is little sense of the power of Moore's Law. The new iPhone is twice as fast as the old one, but shaving fractions of a second off processing time is barely noticeable. Users are more interested in additional capabilities than marginal increases in speed.

To understand the true impact of Moore's Law, you have to look back over a longer timescale. Let's take a quick look at three key areas – computing, navigation, and video cameras – and see how they've changes over a couple of decades. The results may surprise you.

Laptop Revolution

In the early 1990s I was involved in purchasing laptops for oil company executives. They wanted something as powerful as their desktop machines, a real issue at a time when spreadsheets took minutes to complete recalculations. Previous computers took strong arms to lug around, and some of them were merely "portable"' rather than "mobile" and required a main power source. The displays were crude and they were nowhere near as powerful as the desktops of the time.

We solved the problem with the Compaq LTE/386, a novel design, which folded the motherboard into a minimal space. It weighed just six pounds and was slim enough to fit into a briefcase. It was so small that a new name was invented: the tiny laptops were known as Notebook PCs.

For the first time, a laptop could do everything that a desktop PC could do. It displayed the results on a crisp 9.5" monochrome screen (resolution 640 x 480), and battery life was an impressive three hours. A miniature modem inside the laptop allowed the exec on the move to plug into a phone jack and access emails at a speed of 14k bits per second – about .01 mbs in modern terms. All this for a mere $6,600 ($11,400 in 2015).

At the same time, lower-graded staff received the bulky "lunchbox" portable SLT/386 computer. This cheaper machine was similarly powerful but weighed in at fourteen pounds and was not nearly as appealing.

The new Notebook PCs were enviable executive arm-candy. They may have inspired some of the managers – the sort who had a secretary print out their emails so they could read them – to become computer-literate. These days a ten-year-old wouldn't accept one as a gift.

The old Compaq laptop is not only outperformed by any $400 laptop today, but also by tablets and smartphones. The least of these packs far more computing power than the old Compaq LTE. If you're on the move and want to check your email, edit a document, or look at a spreadsheet, you can do it on your phone. "Excel in bed," suggested a recent Nokia advert, headlining their phone's ability to run spreadsheets anywhere.

The raw numbers starkly illustrate how modern machines outperform their forebears. The LTE/386 from 1991 had a processing speed of 11 million instructions per second or MIPS. This made it several hundred times faster than the seventy-pound Apollo Guidance Computer used in manned space landings.

The iPhone 6 is rated at over 25,000 MIPS, making it over two thousand times as fast as the LTE laptop. The iPhone's storage space of 64 GB is more than a thousand times as great as the laptop's 60 megabyte hard disk. Communications are more than a thousand times as fast – and wireless. The screen has three times as many pixels -- and it is in color. It also refreshes faster. Battery life is perhaps the only aspect that has not improved greatly, but you can still expect eight hours of use.

It is as though the F-35 was a tenth the size of an F-16 and a tenth of the cost.

Video Cameras

The evolution of video cameras has been just as dramatic, boosted largely by digital technology taking over from analog.

"Truly amazing," wonders Doc Brown in the 1985 movie *Back to the Future*, admiring a video camera transported back in time to 1955. "A portable television studio."

The machine impressing Doc is a JVC GR-C1U, a camcorder the size of a loaf of bread weighing four pounds. It records twenty minutes of video on VHS tape. It has a black-and-white display and a x6 optical zoom and cost around $1,500 ($4,200 in 2014 money). When you compare it to the thirty-pound camera on the QH-50 DASH – state of the art in the movie's 1955 setting – the 1985 camcorder is a technological marvel. But the technology has been dwarfed by what followed.

Skipping ahead thirty years – as Marty McFly later does – we have the Samsung IIMX-F90 video camera. For about $200, you get a device that weighs half a pound, with a more powerful zoom (52x) and which records nine times as much footage at higher resolution (HD, 1280 x 720) than Marty McFly's JVC. Being digital does not just mean that you can grab single frames as photos. It also means that the image is stabilized and sharpened by a powerful on-board processor.

However, not so many people have camcorders these days, as they already have smartphones that can shoot video. The iSight camera on an iPhone 6 outperforms Marty McFly's old JVC, the camera itself weighing a fraction of an ounce. Being fully integrated with the smartphone gives access to a range of editing, titling, and special-effects software. It's not just a television studio anymore; it's an editing suite and production company too. And thanks to integration with the internet and YouTube, it's a television station that can put your video out to billions of viewers.

Navigation

Satellite navigation was revolutionary when it was introduced a few decades ago. A constellation of US military NavStar satellites sends out precise radio time signals; by locking in to these signals, a user on the ground can calculate where they are in relation to the satellites and hence their position on the globe. It was originally intended for vehicles – ships and aircraft -- and cruise missiles, but when the technology was small enough it became portable.

The original AN/PSN-8 portable GPS made by Rockwell Collins in 1988 was a backpack unit weighing seventeen pounds and costing an eye-watering $45,000 ($90,000 in 2014). It could only find one satellite at a time, and acquiring three satellites and calculating a location took around five minutes. On top of this, the disposable batteries weighed half a pound each, and the backpack unit burned through them rapidly.

Flawed as it may sound now, the new GPS seemed miraculous, and soldiers loved it. An old Army joke has it that, "The most dangerous thing in the world is a Second Lieutenant with a map and a compass," and units getting lost or turning up in the wrong place were a fact of life. Less amusing were air strikes or artillery barrages hitting completely the wrong area. GPS meant you could get precise map-coordinate anywhere in the world, without taking compass bearings or using landmarks, even in pitch darkness, driving rain, or a sandstorm.

You literally could not get lost, anywhere on the surface of the Earth. And operator could call down artillery fire or air strikes with pinpoint precision, in theory without any risk of error.

Within four years the Army had ordered 75,000 portable GPS units, with subsequent iterations after the AN/PSN-8 dropping from seventeen to ten pounds and then four.

Such technology was initially confined to the military, but it was followed by commercial versions for surveyors and others. In the early 90s, handheld GPS units were available for consumers. They were slow and expensive and did not show a map, just grid coordinates. They were much in demand in the 1991 Gulf War among troops when the Pentagon's procurement process failed to deliver the military version in adequate numbers.

Since then the technology for satellite navigation has shrunk to chip size. If you want a standalone unit, you can get one for less than a thousandth the price of the original Rockwell Collins. Instead of a backpack, it fits in your pocket. It gets a precise location in seconds rather than minutes. Like the camera, it's been incorporated into digital technology, so rather than giving you coordinates to look up on a paper map, you get a moving color map display. And its computer can calculate the route for you and guide you every step of the way.

While many people have a SatNav in their car, many have the benefit of GPS without a specialized device. Like the digital video camera, satellite navigation is one of the many functions built into their smartphone.

Putting it all together

This convergence has given ordinary smartphones computing, navigating, and imaging capabilities that surpass anything available a decade or two earlier at a fraction of the cost. The point here is that these are all technologies important for military drones.

In the past, the military were way ahead in high-tech electronics, and there were no civilian equivalents of what they were using. Everything was a matter of custom-made chips with unique capabilities. But now the "big green boxes" of military electronics are increasingly being outpaced by their civilian counterparts from the world of smartphones.

Back in the days of the Compaq LTE laptop, the British military developed a shoebox-sized supercomputer called Mousetrap with custom electronics (8). This ran at 180 MIPS, ten times as fast as the best laptop. It could look at a 3-D map of terrain and calculate lines of sight, so a patrol would know where they needed to go to stay out of sight, or what hills would give a good view of a given spot. This was the sort of thing only the military could do. But because the generation time for military electronics is six years or more, commercial electronics are now way faster than their military counterparts (remember the 25,000 MIPS iPhone). The current radar in the F-15, just nine years old, is geriatric in terms of consumer electronics.

This is hardly surprising when you look at the numbers. By one estimate, the smartphone industry's research budget was around $150 billion in 2014 (6). This dwarfs the Pentagons entire R&D spend of around $60 billion for everything from submarines to satellites and all that lies in between. In aviation, the Pentagon is the biggest player in the game; only a handful of other countries can afford to develop fourth and fifth-generation combat aircraft like the F-22. But when it comes to small electronics, the Pentagon does not have the budget to create anything as slick and sophisticated as the latest smartphone.

At the start of the Iraq war, considerable effort went into developing handheld translation devices with names like Phraselator and Squid. These store phrases in a local language – Arabic, Iraqi, Pashto, Urdu, Dari, and Kurmanj – and play them on command. By 2009 a third generation of device was ready, with the rugged handheld Phraselator weighing "only" 13.5 ounces (three times as heavy as the iPhone 3 launched in the same year). By 2011 the Pentagon wanted something more sophisticated and had a $50 million budget – just as an app called SpeechTrans came along, doings pretty much what they wanted on a commercial phone. In late 2011 Raytheon demonstrated a "militarized" translation app which ran on an Android tablet. Dedicated hardware for translation was looking like a thing of the past.

Perhaps the most conspicuous case of military electronics getting outclassed came with "Ground Soldier System Increment 1," a program to equip troops with networked tactical computers including navigation, communication, and data-sharing capability. It is also known as Nett Warrior after WWII hero Robert B. Nett. The initial version weighed over seven pounds and was universally unpopular.(9)

In 2013 the US Army unveiled a new lightweight version based on a Samsung Galaxy Note II rather than military electronics (and nicknamed Droid Warrior for its Android operating system). The tablets were bought for $700 each, making them cheaper as well as much lighter than the previous iteration. Commercial, off-the shelf equipment (COTS) is becoming respectable in military circles.

DARPA have even developed an app called Kinetic Integration Lightweight Software Individual Tactical Combat Handheld ("Kilswitch"), which transforms the process of calling in an air strike from a technical specialty to an app that a teenager could master in minutes. All you need is an Android tablet and a connection.(10)

"Big Army is slowly getting away from six-year acquisition cycles to develop a ruggedized piece of hardware that is outdated before we get it to the soldier," one official told National Defense Magazine.

These days many people use a smartphone rather than camcorders or stand-alone GPS, and tasks like checking email which previously required a laptop have been absorbed by phones. These capabilities are highly relevant to small drones.

As we saw in Chapter 1, in 1944 getting a television camera small enough for the TDR-1 drone was a technical challenge. Now miniature video cameras, complete with everything needed to encode and transmit a moving image via radio signal, are a cheap commodity item. Navigation for the ill-fated Aquila drone relied on an expensive, purpose-built inertial sensor. You can now get a more capable navigation system based on GPS off the shelf for a few dollars. The computing power needed for the maneuvers of a Firebee is no longer an expensive, dedicated piece of electronics; it's already on your smartphone and easy to program. And if you need sensors for acceleration, gyroscopes to detect tilt, or even a barometer to measure changes in altitude, they are already there.

So why not use smartphone technology to build a cheap drone?

The Men from MITRE

Michael Balazs and Jonathan Rotner from US government think-tank MITRE Corporation are an energetic and engaging double act, constantly bouncing ideas off each other. They aim to show the Pentagon that costs for military-grade small drones can be slashed by an order of magnitude, and with the slogan "90% of the Solution for 10% of the Price" they set out to exploit the power of technology developed for the smartphone industry. (11)

Small military drone developers have always insisted that their products cannot follow Moore's Law. They are dealing with custom hardware, so the volumes are small. Electronics have to be constructed to meet military standards, with tight requirements for tolerating temperatures and shock. Everything has to be tested and certified.

"We haven't seen Moore's Law apply to small UAS [Unmanned Aircraft Systems] in the same way it applied to computers. This is because UASs are multifaceted devices that rely on advances in computers and avionics, materials, batteries, and other disciplines to evolve," one drone developer told me.

Balazs and Rotner challenge this line of thinking. MITRE is not a manufacturer and does not compete with the drone-making industry. Its role is to provide the government with solutions, and the MITRE developers are talking to industry and government – and anyone else who will listen – about translating their ideas into products.

It started in 2012 when Balazs was rebuilding an old military robot, like the remote-control robots the Army uses for bomb disposal. The project was a tricky one because of the outdated proprietary electronics and custom software involved. It was difficult to gets parts, and difficult to program the machine when you did.

This kicked off a conversation about standardizing military robots using off-the-shelf components and open-source software.

"We took a step back and looked at the leading technology – smartphones," says Balazs. (12)

Android phones are powerful and cheap. It's also easy to connect them to other electronics by adding a simple "breakout board." Much of the necessary functionality for a drone, including GPS navigation, video camera, and radio communications, is already built into the phone.

Rotner describes their approach as "the path of least resistance" as it was so easy to get the right hardware from commercial sources.

Balazs and Rotner worked with a team from the University of Virginia to demonstrate their Android Control and Sensor System (ACSS) with radio-controlled toy cars. Their team of robot cars explored a sports stadium without external control, using instruments to build up a map of the temperature and sunlight intensity. The robots communicated with each other, formed a network, and carried out coordinated tasks, all for just a few hundred dollars.

This is no small feat. The ACSS machines were demonstrating more autonomous intelligence than most military ground robots or UGVs [Unmanned Ground Vehicles]. And as one commentator pointed out, typical military robots cost more than a luxury Mercedes "and you don't even get leather seats." (13)

Programming was easy because Android is an open operating system, with hundreds of thousands of applications available. This gave the team access to a treasure trove of software that could be downloaded for pennies. Getting enhancements proved easy, and in some cases free. Rotner mentions an app they downloaded for streaming video that lacked a remote on/off command. They contacted the developers, and within twenty-four hours an improved version with the new control was available at the app store at no extra cost.

While programming the embedded electronics on most robotics is a specialized skill, anyone who knows the common programming language Java can design Android apps. The team took advantahe of existing computer vision software to help develop an app to follow a road autonomously. Similar developments help with avoiding obstacles or recognizing particular objects. Any clever algorithms or technique developed in the industry can be shared thanks to the open source approach.

Having proved the principle by building Android ground robots, the MITRE team set out to build an Android-based, military-grade drone. Finding that Raven-type hardware was prohibitively expensive, they decided to make their own with a 3-D printer. (14)

They were not the first in this field. In 2011, Andy Keane and Jim Scanlan of the University of Southampton computer-designed a drone with a five-foot wingspan and printed the parts in hard nylon on a 3-D printer. This is a device which builds up a solid object layer by layer, printing it a fraction of a millimeter at a time. The process is sometime called 'additive manufacturing.' You can see this in the finished wings, which have faint marks like the rings of tree growth. Since then, drone designers and enthusiasts have been made wings, fuselages, and other components, and printing entire aircraft is now routine. The University of Sheffield's Advanced Manufacturing Research Centre is building a reputation in this area and now optimize drone designs so they can be printed out quickly on 3-D printers with as little waste as possible.

There is a whole community, known as the Maker Movement since 2005, based around sharing designs and making them with 3-D printers, and this includes drones. Spanish company CATUAV offers a one-week course in 3-D printing drones; Ang Cui, a Columbia University PhD, has a Drones at Home blog with a step-by-step guide for would-be drone makers to follow.

MITRE's first printed aircraft took a week to print, comprising some seventeen sections. It crashed when a cross-wind caught it while landing in Colorado – Balazs and Rotner act out the crash in their live demonstrations – but this gave the team the chance to see which pieces broke and which survived. They carried out a swift redesign and sent the modified CAD files back to base in Virginia. By the time they returned to the laboratory, the improved version had been printed out and was ready to assemble.

The first-generation drone proved the principle and cost around $6,500 to put together. Razor is the second-generation model, optimized for 3-D printing, and it can be printed in a day for about $550. Razor has a wingspan of forty inches and can cruise at 45 mph for forty minutes.

Razor's airframe is printed from tough Ultem plastic. The team experimented with a vast range of different printers and materials – they say they tried everything except food printers, in fact, and they wouldn't mind trying those. Razor is held together by small plastic X-connectors. In a crash, these give way first, leaving the wing sections undamaged, similar to the way the Raven "disassembles" during a hard landing. A handful of spare X-connectors is all the repair needed for most accidents. Add the engine, batteries, Android phone, and a miniature autopilot and the total cost of this highly capable and adaptable drone is less than $2,000.

Rotner says 3-D printing embodies the same flexible approach as using Android apps. Designs can be shared freely and benefit from enhancements by a large pool of developers. Changes can be put into production instantly. It is simple to redesign the payload bay or any other element to meet new requirements. For example, Balazs says they have experimented with alternate sets of wings optimized for long endurance or high speed (over 100 mph), which can be swapped out as needed. Instead of a one-size-fits-all approach, every customer can have an aircraft precisely tailored to their requirements at no extra cost.

To show what Razor can do, the team used it to take aerial pictures of the landscape beneath and then automatically stich them together into a mosaic map. This is like the Google Earth view, except in higher resolution and showing everything as it is now. An instant bird's eye view of the territory below, exactly what a military commander might want to see, but also useful for agriculture and commercial purposes.

Powerful Android phone processors carry out processing on board. Instead of beaming back gigabytes of video to a ground-control unit for analysis, the drone reports back only when it identifies something of interest. This is important because of bandwidth limitations that will apply increasingly as the skies get crowded with drones. The on-board processing is carried out using GPUs, the smartphone's graphical processing units developed mainly for games. These were initially very expensive, but decades of development in the industry has seen GPUs follow the familiar curve of Moore's Law.

Razor's military-grade autopilot is currently a separate unit costing a few hundred dollars. Better ones for small drones can cost

$6,000, but the team would like to see an equivalent autopilot as an Android app. The issue is caused by the fact that Android is multitasking rather than responding in real time. Dedicated autopilot computers currently have the edge, but it's only a matter of time before smartphones overtake them and the dedicated autopilot goes the way of the dedicated translator.

The MITRE team have sketched out plans for a smaller, portable tactical drone costing less than $1,000. Prices may fall even further. 3-D printing is in its infancy, but prices will drop as the mass market picks up, economies of scale kick in, and new technology evolves – and Moore's Law strikes again. Around 35,000 3D printers were sold in 2012 according to tech analysts Gartmore; in 2015 they estimated 250,000.

These printers are improving rapidly. In late 2014, Hewlett Packard introduced a new industrial 3-D printer the size of a washing machine which prints ten times as quickly as the previous model and costs about half as much. (15) This is following a similar pattern to laser printers; the original HP LaserJet sold for $3500 in 1984 ($8,000 in 2014 dollars) and weighed seventy pounds. The modern equivalent is a fraction of the weight, prints three times as fast, and costs less than $100. This suggests that the cost of producing the airframe is likely to come down.

Similarly, the cost of Android phones is steadily declining and there is now no shortage of phones at retailing at $50 and less.

Cheap phones will feature in another MITRE scheme, a drone with multiple fixed phone cameras embedded in the wings and fuselage. The intention is for images from the cameras to be fused together to create a single wide-angle, high-resolution image of the ground below. Rather than sending the whole video back as a stream and eating up bandwidth, the operator could zoom seamlessly between the big picture and a small portion at higher resolution.

Powerful Android processors have also had the team looking at pushing the boundaries of autonomy. The usual approach puts processing power in the ground control station, but with Razor you don't necessarily need one. The drone has everything it needs to carry out a mission without any input from the ground, and it can share video or other data with anyone who has appropriate communications and a display screen. Razor has the potential to be a very smart, autonomous drone. We will look more at what this will mean in Chapter 7 on swarming.

The commonality with smartphone technology had another bonus. While people in the industry and the military may struggle with some of the details of embedded firmware, Balazs and Rotner find that their audiences, including eight-year-old schoolchildren, easily grasp the idea of downloading software in the form of new mission apps from an app store. Their design is simply a flying smartphone, an easy idea to understand.

The most conspicuous deficiency of the Razor is an infrared thermal camera for night operations. Such cameras can cost thousands of dollars and may be the most expensive single component on a drone. But, as if on cue, a consumer thermal imager, the FLIR ONE, was launched in 2014. (17) This is made by FLIR, a leading supplier of thermal imagers for the US military. The FLIR ONE, designed as an add-on to the iPhone, has low resolution (80 x 60 pixels, converted into 160 x 120 by the software), which is much less than military-grade devices. But it retails for just $250, and provides a useful infrared capability. A drone with this type of sensor can spot warm bodies in pitch darkness.

Competition for the FLIR ONE arrived quickly. Raytheon's Seek Thermal camera, also an add-on for iPhone or Android smartphones, has a resolution of 206 x 156 and comes in standard and extended range versions, all for $50 less than FLIR ONE. (18) The Seek Thermal is claimed to be able to detect human heat signatures from a thousand feet away. Future generations of imagers are likely to follow the path of GPS, video cameras, and computers: ultra-cheap, ultra-light, and possibly built into most smartphones.

Phones in Space

This idea of smartphones being used for unmanned vehicles has cropped up elsewhere. Satellites have always been unmanned. Although rarely described as such, they are effectively unmanned spacecraft. Again developers are experimenting with smartphones as a replacement for expensive dedicated electronics.

Surrey Satellite Technology Ltd is a British company that specializes in low-cost satellites to make space more accessible. Stuart Eves, SSTL's Principal Engineer, points out that between

1999 and 2008, imaging satellites for surveying dropped from the size of a pickup truck to the size of a chest freezer. (19)

"It sounds like magic," says Eves. "But it's just technology progressing."(20)

A key aspect of SSTL's approach is using commercial processors and other off-the-shelf components rather than those custom-built for the space industry. Eves says that by the time a bespoke processor has been developed for a satellite, technology has moved on another generation. The ultimate in off-the-shelf technology is the smartphone, and in February 2013, a ten-pound smartphone satellite developed by SSTL and Surrey University blasted off into space. This was STRaND-1 (Surrey Training, Research and Nanosatellite Demonstrator 1), first of its kind.

There are many other phonesats under development, and these are merging with the new standard for "cubesats," satellites that fit a standard format of a cube, four inches on a side. NASA has a whole PhoneSat program, with the latest launch being PhoneSat 2.5 in April 2014. PhoneSat no longer uses entire phones, but is based entirely on smartphone technology.

"We leverage all the transistors and the radio components and the amps and the resistors, we leverage all the developments that are happening. We take the latest CPUs [central processing units], the latest flash drives, the latest sensor systems and stuff them into our little box," developer Will Marshall told Aviation Week. (21)

At present, these tiny satellites are research tools. They prove that phone electronics can survive in space and are capable of controlling satellite operations; PhoneSat 2.4 used a magnetometer (the smartphone version of a compass) to align the craft. But applications are already being found for phonesats with simple instruments. Large numbers of cheap satellites would provide a good way of monitoring "space weather" , the geogmanetic storms and flux of charged particles over a broad expanse of space. Arrays of them may be able to function as space telescopes or communications platforms.

The PhoneSat experience reinforces the MITRE team's findings: consumer electronics really are powerful, reliable, and rugged enough to carry out tasks in the real world, controlling a ground robot, a drone, or a satellite.

Android Invasion

Military developers are also latching on to the potential of commercial electronics. DARPA has noted that remote sensors used to detect passing vehicles have the same functionality as smartphones. Their core requirements – "processing, storage, communications, navigation, and orientation"– are similar to what smartphones do. So their program, known as ADAPT ("ADAPTable sensor system"), sets out to use consumer electronics based on the Android operating system.

Like MITRE, the DARPA researchers have adopted an open-source software framework, so anyone can create, refine, and distribute new software. The commercial Android core is far cheaper than the equivalent bespoke military device, and cheaper to program.

ADAPT provides a standard core which can be plugged into all sorts of devices, not just sensors. As a demonstration, ADAPT core was plugged into a small quadrotor drone as a controller. The core, with it processors, memory, communications and connectivity, is a step towards a universal "robot brain on a chip" that can be plugged into any body as required.(22)

The Navy is also waking up to the benefits of open architecture, using modular hardware and standard software for a range of robots. The Advanced EOD [Explosive Ordnance Disposal] Robotic System is their test program for using off-the-shelf technology rather than getting locked into specific vendor products. The developers believe it will bring the usual open-architecture benefits: more affordable, easier to support and upgrade, and being modular, new components can be added more easily. Rather than one machine, the Advanced EOD System will be a whole family, all with the same brains but different size bodies from the man-portable to golf cart. If successful, this will lead to the Navy towards the open-source paradigm.

Flying Circuit Boards

Seen from the money-no-object view of military aviation, a few thousand dollars for a drone is cheap. But to an impoverished academic researcher, it's still expensive. In 2009, three Swedish consultants decided to try and build the lowest-cost possible drone. Their concept was a "flying circuit board" with four rotors attached.

They set up a company, Bitcraze, to produce the craft called Crazyflie. It now sells for $149 including $33 for the controller. Crazyflie weighs less than an ounce and fits in the palm of your hand. The little quadrotor flies for about seven minutes on one charge. But it is a genuine drone, intended as a platform for researchers exploring flight control and autonomous flight.(23)

Crazyflie is of course open-source, and numerous "hacks" are published on Bitcraze's web site. Some of these are hardware hacks, like buffers to protect the rotor from impacts. Most are pure software, things like automatic stability recovery or software to flip over in mid-air or to perform other unlikely maneuvers. Some developers are looking at optical control, tracking and steering the drone's flight with remote sensors, or automatic navigation.

But even Crazyflie can function as a practical tool. Researchers at Michigan Tech Research Institute have been using the miniature drone to explore tight spaces such as the interior of fuel tanks that are inaccessible to human technicians.

Miniature helicopters have already had their debut on the battlefield; the British Army started using PD-100 Black Hornet drones in Afghanistan. Like the Crazyflie, this is a helicopter you can hold in the palm of your hand; the carrycase with the controller and two air vehicles fits into a military pocket.

The Black Hornet is produced by a Norwegian company, Prox Dynamics. What makes it remarkable is the design which allows it to fly in a ten mile per hour wind. The company previously developed a passive stability control technique for miniature helicopters, now widely used in radio-controlled toys. It may tumble and flip over in flight, but it recovers and continues the mission. Like the Raven, it navigates via GPS and has a handheld controller with an intuitive interface and beams back both day and night video.

Prox Dynamics has a slew of patents protecting the various ingenious design features that distinguish the Black Hornet from its competitors. Black Hornet currently costs about $150,000 for a controller and two vehicles, making it hundreds times as expensive as Crazyflie. The question is whether the advance of technology, and the army of developers hacking for Crazyflie, will produce anything as capable as Black Hornet. The patent situation will force them to find novel solutions, and judging by history, it is only a matter of time.

Drones Everywhere

Razor is certainly not Raven, but looks more advanced than the previous-generation Pointer. Future low-cost craft will be better thanks to Moore's Law. Small drones are currently at the stage of the Compaq LTE, Marty McFly's JVC, or the original backpack GPS: they look impressive because we've never seen anything like them before. But they are likely to look feeble compared to the generations to follow.

Small drones are improving rapidly in different ways, as we will see over the next four chapters. We know already that the basic capability of getting eyes on a target need not cost tens of thousands of dollars. Augustine's Law does not apply to small military drones. The upward cost spiral of Augustine's Law is trumped by the downward spiral of Moore's Law.

It is not just about putting eyes on a target. The cost of putting a grenade on target in the manner of LMAMS or Switchblade is also becoming affordable. The world's smartest weapon will cost a few thousand dollars or less and can be made anywhere by anyone. Unlike stealth, this does not require classified technology, and unlike the F-22 or F-35, it does not require a research and development budget of billions or even millions. You cannot destroy the factory because the factory is any 3-D printer.

As we have seen, it is hard to put an exact price on an F-35, but at a median estimate of $150 million, one costs the same as about seventy-five thousand Razor drones at $2,000. In reality the drones should be far cheaper, especially when bought wholesale rather than retail. (The iPhone 6, which retails at over $600, cost around $200 to produce in China, a pattern repeated with other high-end smartphones).

MITRE started their project back 2013. Electronics are cheaper now and will be even cheaper in years to come. There are already $50 smartphones with GPS, camera, and a more powerful processor than the original iPhone. The Samsung Solstice is $35 on Amazon at the time of writing. Moore's Law is still in force.

If drones cost $2,000, a squadron of sixteen F-35s costs the same as about a million small drones. Even the standard AIM-120 AMRAAM (Advanced Medium-Range Air-to-Air Missile) used in air-to-air combat costs as much as several hundred drones.

An F-35 squadron might annihilate a few dozen cheaper fighters but is helpless against such vast numbers. We will be looking into exactly why in a later chapter.

Small drones still resemble battery-powered toys in that they run out of juice after an hour or so, and even large numbers of small drones cannot take over the missions of traditional airpower. In the next chapter, we will see how that apparent limitation is being shattered by small drones that can fly forever.

REFERENCES

1) "The Lehmann 960, for example…"

Drones L-M Series. (n.d.). Retrieved from
(http://www.lehmannaviation.com/l-m-series.html)

2) Augustine's Law

Smallwood, D. (2012). Augustine's law revisited. *Sound and Vibration*. (http://bit.ly/1N5QzVU)

3) "Whatever happened to the lightweight, low-cost fighter we signed up to buy?"

Burton, J. (1983). *The Pentagon Wars: Reformers Challenge the Old Guard*. Annapolis: Naval Institute Press

4) F-22 Cost

General Accounting Office. (1998). *Aircraft development - The Advanced Tactical Fighter's costs, schedule, performance goals*. (Report number: B-229129). (http://1.usa.gov/1OA8MPS)

5) F-22 superiority

Kopp, Carlo. "Just How Good Is the F-22 Raptor?" *Air Power Australia*. (http://bit.ly/1MYFNTe)

6) Smartphone R&D Budget

Samsung breaks R&D spending records, invests $14 billion in new tech. (2015, March 10). (http://bit.ly/1OA8XL3)

7) "The $659 Ashtray"

Connell, J. (1986). *The New Maginot line*. London: Coronet Books

8) Mousetrap computer

Harley, R., Palmer, K., & H.C., W. (1991). Mousetrap, a miniaturised supernode for terrain modelling. In *Transputing '91: Proceedings of the World Transputer User Group*. (http://bit.ly/1NsBaSV)

9) Nett Warrior to Android Warrior

Dixon, Alex. "Nett Warrior gets new end-user device." *WWW.ARMY.MIL.*

10) "Kilswitch"

Lamothe, D. (n.d.). Infantry overhaul: How DARPA's new experiments could shake up ground warfare. In *Washington Post.* (http://1.usa.gov/1PAsFb1)

11) MITRE and Razor

Balazs, Michael and Rotner, Jonathan, "Inexpensive, Android-Controlled ISR Solutions", MITRE paper MP120729, January 2013

"Shaving Costs." *The Economist.* (http://econ.st/1kYV4bf)

12) "We took a step back and looked at the leading technology"

Michael Balazs interview with the author, November 2013

13) "And you don't even get leather seats"

William, W. (2013, November 15). UGV cost & why they should have leather seats. (http://bit.ly/1LRdlyR)

14) Keane & Scanlan First 3-D Printed Plane

3D printing: The world's first printed plane. (2011, July 27). (http://bit.ly/1IvRaNm)

15) New HP Industrial 3-D printer

Mearian, L. (2014, November 7). HP's new 3-D printer is aimed at manufacturing, not consumers. (http://bit.ly/1u8qAa3)

16) FLIR ONE (http://bit.ly/1d7MEoz)

17) Seek Thermal

18) Surrey Small Satellites

Hambling, D. (2009). The big promise of small satellites. (http://bit.ly/1XCnknS)

19) "It sounds like magic."

Stuart Eves interview with the author, October 2013

20) "We leverage all the transistors and the radio components"

Morring, F. (2014, August 6). Smartphone advances drive smallsats. (http://bit.ly/V1Jcdc)

21) DARPA ADAPT

Lawrence, C. (n.d.). ADAPTable Sensor System (ADAPT). (http://bit.ly/1jB3aI8)

22) Bitcraze Crazyflie

The Crazyflie Nano Quadcopter. (n.d.). *Bitcraze Crazyflie.* (http://bit.ly/1TnuVjI)

23) Prox Dynamics PD-100 Black Hornet (http://bit.ly/1IIGIrf)

CHAPTER FIVE

THE FINE ART OF FLYING FOREVER

> *"Once you have tasted flight you will walk the earth with your eyes turned skywards... there you will long to return."*
>
> - Leonardo da Vinci (attributed)

There is an old, pre-smartphone-era joke about a man who meets his inventor friend carrying two suitcases down the street. The inventor shows off the wristwatch he has built, which as well as indicating the time anywhere in the world, also shows the tides, phases of the moon, and position of the stars. His friend is duly impressed. The only problem, says the inventor, picking up the suitcases, is the size of the batteries.

That about describes the present situation of small drones. What they do is great; the only problem is the batteries. Whatever small drones can do, they cannot do it for long because of their limited battery life. The smallest only fly for ten minutes or so. Quadrotors may manage half an hour, and the Raven, our gold standard for capability, keeps flying for about ninety minutes.

The situation is better the larger the drone. The RQ-20 Puma, the Raven's big brother with twice the wingspan and three times the weight, has an endurance of three hours. Scaling means that, other things being equal, bigger drones carry proportionately more batteries or fuel.

Switching from battery power to liquid fuel extends the range of small drones dramatically. In 2003, a drone weighing eleven pounds with a wingspan of six feet flew across the Atlantic, from Cape Spear in Newfoundland to Mannin Beach in Ireland.(1) The drone, known as TAM 5, took thirty-eight hours to make a crossing of almost two thousand miles, at a steady speed of around fifty miles an hour. Small drones can cross oceans and continents.

Unfortunately, the internal combustion engine has disadvantages. For one thing, it is very noisy; this was one of the flaws that sunk the RQ-16 T-Hawk. It cannot sneak up and take pictures without being spotted.

A drone with a combustion engine is more vulnerable to being shot at, and when it crashes, there is a safety hazard. Those falling-apart landings, or running into a wall, become less hilarious. And the logistical issues – having to keep stores of drone fuel and drone oil, servicing and maintaining the engine – make it more complex and less attractive than the battery-powered alternative.

On the face of it, small drones are doomed to perpetual disadvantage compared to their larger cousins. The Raven cannot match a Predator's ability to orbit over a target area for twenty-four hours or more, and the Raven is confined to a small area. With current technology, a Raven could in theory fly out thirty miles, take a picture, and then return to base before the batteries gave out, but that is literally as far as it goes. In practice, because a battery reserve is needed and there is likely to be wind, the maximum range for a Raven is much less.

One way of dealing with this is a drone carrier. A transport aircraft, manned or unmanned, acts as a mothership, carrying a fleet of drones to the battle area. It then releases them to fly the last few miles under their own steam. For expendable, Switchblade-style drones this would be a one-way trip, as they are effectively miniature cruise missiles. Reconnaissance drones could return and be picked up by the mothership via a net, just like the old Firebees. The drone carrier, like the aircraft carrier, could become a powerful tool for force projection.

However, it would be wrong to assume that small drones could never carry out strategic reconnaissance or strike missions. In the near future, small drones will stay on the wing 24/7, with mission times measured in days, months, or longer. A clutch of different technologies are contributing to this transformation.

Batteries That Are Stronger, Longer

Over the years we have seen batteries progress from lead-acid to nickel-cadmium (NiCad), nickel metal hydride, lithium-ion, and the related lithium phosphate and lithium polymer cells.

Years of intensive research have pushed these types of cells close to the theoretical maximum for the battery chemistry involved. The current state of the art are the batteries in the Tesla Model S, packing in about 240 watt-hours per kilogram; each pound of battery holds enough energy to boil about a third of a pint of water. (2) That is less than a tenth of the energy density of gasoline or other liquid fuel, but several times greater than NiCads.

Drones like the Raven have the same rechargeable lithium-ion battery technology as laptops, phones, and electric cars. Any improvement will mean a change in battery technology.

Lithium-air batteries look like an attractive alternative to lithium-ion. They take oxygen from the air and have many times as much power per pound. In fact, the energy density of lithium-air batteries is close to that of gasoline. The problem is making them safe, as lithium-air cells produce a lot of heat during operation. This is hardly surprising; the chemical reaction involved is similar to burning highly reactive metal. The designers also need to find a way of avoiding contact between the metallic lithium and water in the air, otherwise spontaneous fires can result.

IBM gave up on lithium-air battery development in 2014 after five years of work. The Joint Center for Energy Storage Research, set up by the Department of Energy, did the same after two years. Both concluded that while lithium-air might be the future in the long-term, it was not yet feasible. (3)

A more promising variation is the "molten air battery" created by researchers at George Washington University with support from the National Science Foundation. This features a molten electrolyte and takes oxygen from the air like lithium-air. Its electrodes are made of special materials such as vanadium boride, which can handle more electrons per molecule than lithium. Researchers estimate that this technology will be able to store twenty to fifty times as much energy as lithium ion, but it is likely to take at least a decade to develop. (4)

There is a gentler and more immediate alternative in the form of lithium-sulfur (Li-S) chemistry. This stores less energy than lithium-air but far more than lithium-ion, and it is also significantly cheaper. The main component, sulfur, is a by-product of the oil industry. Oil refineries produce millions of tons of sulfur every year, whereas Li-ion batteries require expensive metals like manganese, nickel, and cobalt. Lithium-sulfur batteries are also safer and more environmentally friendly than lithium-ion. (5)

Researchers around the world have been working on Li-S for years, but have struggled with a couple of basic problems. When the battery is discharged, a layer of lithium sulphide tends to build up around the anode, degrading the cell capacity after a few cycles. The other problem is that the conversion to lithium sulphide and back involves swelling and contraction – the bulk can change by more than half – putting mechanical strain on the cell.

However, researchers now claim to have overcome these problems. The British company OXIS Energy aims to be first to market in 2016 with a lithium-sulfur battery. The Li-S cell consists of a lithium metal anode and a sulfur-based cathode in a polymer binder, separated by an electrolyte.

Lithium batteries carry an element of risk because of the growth of root-like metallic lithium "dendrites" within the cell. These can cause short circuits resulting in overheating and fires. However, the sulfur electrolyte in Li-S creates an insulating layer of lithium sulphide that nullifies this risk. Further, OXIS' electrolyte is non-flammable. OXIS batteries have passed a wide range of abuse tests including overcharging and short-circuiting. The British Ministry of Defence has tested safety by firing bullets through them, and even then the cells continued to function safely, though at reduced capacity.

The initial commercial version already stores almost 50% more energy per pound than lithium-ion, and that will to improve. OXIS has a roadmap that includes reaching 500 Wh/Kg by 2018 (more than double the Tesla battery), and they are well on the way with prototype cells. Looking further ahead this should double again as the full potential of lithium-sulfur is developed as Li-ion has been. They also benefit from an unlimited shelf life, as lithium-sulfur cells do not require periodic recharging like Li-ion.

Prototype lithium-sulfur cells have already been used in experimental drones. Simply upgrading the batteries to Li-S could increase the flying time of the Raven to around six hours in a few years' time. As other battery technology starts this could be extended further.

Fuel cells

Fuel cells are another possibility for power storage in drones and other electronic devices. These generate electric current directly by a chemical reaction. Unlike an internal combustion engine, a fuel cell is virtually silent. The D245XR solid oxide fuel cell, made by Ultra Electronics is one of the smallest fuel cells available. It weighs five pounds and burns propane fuel at eight hundred degrees centigrade. The exhaust is warm, but the fuel cell itself is so well insulated that it can be fitted into a device without heating the electronics around it. (6)

The D245XR is the heart of a new version of the Lockheed Martin Stalker Drone test-flown at the US Marine Corps Warfighting Laboratory in Quantico in 2014. The standard battery-powered Stalker flies for about two hours on one charge; the propane-driven variant, known as the Stalker XE 240, goes for eight hours at a stretch. A follow-on to the XE has been developed with a larger fuel tank, with an endurance of almost thirteen hours.

The Stalker XE runs on normal commercial propane, which it runs through a "scrubber" to remove impurities. Tanks of propane for cooking and heating are available pretty much anywhere in the world. In terms of logistics, it is easier than a gasoline-powered drone.

The Stalker XE is not the only drone with a fuel cell. The US Air Force has been experimenting with an AeroVironment Puma drone similar in size to the Stalker fitted with a Protonex fuel cell running on hydrogen. This has achieved an endurance of seven hours.

The Navy has their own project called Ion Tiger, also using a Protonex hydrogen fuel cell. In one test in 2013, this flew for twenty-three hours; in 2014 this was extended to an astonishing forty-eight hours with the aid of a new cryogenic storage system. Ion Tiger is a bigger craft, weighing some thirty-seven pounds, but it is still an impressive flight time, and it shows the potential of the technology. Where price is no object then this type of technology will find its niche. (7)

The ideal small drone fuel cell can be recharged, converting water into hydrogen and oxygen and back. In early 2015, Boeing announced a project based on just such a fuel cell. A second part of the same project is a solar panel to recharge the fuel cell, a concept which we will explore later in this chapter.

Hanging Around: Perching Drones

Birds have a huge advantage over drones. By landing on a convenient perch, they can keep an eye on the world without having to flap their wings and expend energy.

Multirotor, helicopter-style drones already have some "perch and stare" capability. The Qube, the small quadrotor made by AeroVironment, is designed to set down on any flat surface; long legs give the camera slung underneath enough clearance for a good view. It does need a flat surface though. A drone would have far more flexibility if it could perch on roofs, power lines, trees, or telegraph poles like a bird.

Fixed-wing aircraft have greater speed and better aerodynamics than rotorcraft, but "perch and stare" is harder when you cannot land vertically. As they come in to land, birds open their wings and spread their feathers for maximum braking force to create a controlled stall. They practically come to a complete halt mid-air, making it easy to grab hold of a branch or other perch.

While some hawks land with a hefty thump, an owl can touch down on your wrist so lightly you barely feel it. Other birds use their near-perfect flight control to cope with seemingly impossible perches: parakeets, for example, can grab a vertical branch and hang on. Even something as big and ungainly as a heron can perch with great ease on a rail.

No man-made craft can match this sort of control. Small drones are not, as a rule, fitted with landing gear. They tend to rely, like the Raven, on that decidedly inelegant controlled crash.

The simplest method of perching is to grab hold of something with a hook. This idea comes with a long pedigree: an aircraft landing on an aircraft carrier has an arrester hook that latches on to a cable strung across the carrier's deck. This is nice and simple because the flight dynamics do not demand any fine control. The pilot just has to fly slightly above the line of the cable, and physics takes care of the rest.

The limitation for drones is that a hook only works for cables or wires, and the take-off afterward is as clumsy as the landing. Unless it can take off vertically, the drone must disengage the hook and drop off its perch, gaining speed as it falls, and then it has to pull up sharply or hit the ground.

Aurora Flight Sciences have experimented with this sort of perching for a version of their Skate drone called the Urban Beat Cop. (8) This is intended to remain in an area and inconspicuously keep watch over a long period. Silent and stationary, it would not attract attention, and people are unlikely to notice the drone. This is a clever way of hiding in plain sight; the proliferation of black boxes and other widgets on power lines means we are used to ignoring them.

However, other researchers are more ambitious in their perching goals. Bhargav Gajjar is the founder of Vishwa Robotics in Cambridge, Mass, a company with the motto "Better than Biology." (9) Gajjar found flying small drones entertaining, but the overall experience was frustrating because of the short flight times.

"You fly around for a few minutes and then spend an hour recharging," says Gajjar. "If you could perch you could look around without using any power." (10)

Gajjar wants to exploit the other advantages that perching brings. For one thing, if a perching drone can find a suitable vantage point it can get a more stable, stealthier, and closer viewpoint than it would have circling overhead. Perching would allow operators to carry out the same sort of "pattern of life" analysis that Predator teams can do now, but from tens of feet away rather than an altitude of a mile or more.

In addition to perching, Gajjar also wants his drone to be able to walk, allowing it to explore environments like building interiors or caves. It does not just have claws but actual, bird-like feet and legs.

Gajjar carried out an intensive study of different types of birds and their feet so he could apply their methods to small drones. He took video with a high-speed camera to record their different landings, and found that nature has evolved at least two distinct strategies, one for perches like branches, and another for flat surfaces.

Flight control is crucial for any drone landing on flat surfaces. The drone has to carry out a similar maneuver to a bird, braking sharply and ending with a controlled stall just above the ground. Birds' legs have tendons that act as shock absorbers, and the mechanical version mimics this.

When holding on to a perch, the bird needs a strong gripping action. Gajjar's bird feet are based on a hawk's, with sharp metal claws that can hold on fast. The grip ensures that the drone is anchored at once, so there is no risk of hitting the perch and bouncing off. The strength of the drone's grip is impressive. A video shows a drone with the legs perched on a stick that the handler shakes violently but fails to dislodge. The claws are sharp too; Gajjar says they can easily cut your finger if handled carelessly.

Birds' legs have hundreds of muscles, making them difficult to copy exactly. Gajjar has simplified this to just two motors per leg – one for gripping and one for moving. This is enough for the drone to be able to grip, turn, and walk. The current version can only waddle short distances but does demonstrate the possibility of a drone than can fly into a building though a window, land, and then explore stealthily on foot.

Gajjar produced his legs for the US Air Force under two development contracts. He does not even know which drone the Air Force planned to attach them to; he was just given the specification and delivered two pairs of legs. The Air Force then integrated the legs with their own craft and carried out tests. Gajjar says they were pleased with the results and could not believe the gripping power of the claws. This type of leg would be well suited to something the size of a Raven but could not really be scaled up to something much larger.

Legs are only part of the complete perching solution. Gajjar's design relies on the pilot locating a suitable perch and guiding the drone in; it is a process that calls for some skill, and automated landing would be safer. GPS is far too imprecise for perching, which needs to be calculated to the inch. The next stage will see a drone with its own sensors and enough processing power to carry out the tricky task of identifying a perch for itself.

The Air Force has already awarded a slew of contracts to this end, and as of 2015, there are three competing designs for a perching sensor. These include the Landing Site Assessment, which uses video and other sensors to build a 3-D model of the area to "identify, select, survey, and exploit" possible perches. The ImageNav-LZ system for autonomous landing zone detection has the drone fly in a spiral approach and uses a camera to detect possible perches by the shadows they cast, since these give good three-dimensional information. (11) A third system, ALPS (Autonomously Locating to Perch and Stare), also uses a video feed to assess possible perching sites and then plots an approach to them. Few details of these sensors are available, but all the developers seem confident in their approach.

Less highly classified work is being carried out at the University of Pennsylvania's GRASP Laboratory. Their approach is only for landing on power lines rather than branches or window ledges. Their software scans the drone's field of view for straight lines and then calculates the distance using a technique called optic flow (we will examine this technique in more detail in Chapter 6).

Another approach is to locate a power line by sensing its electric field, rather than by using a camera. This performs two functions at once: the drone can spot and avoid potentially dangerous power lines, and it can also localise them accurately enough to be able to perch on them without visual guidance.

Perching on power lines also opens up another possibility: recharging by stealing electricity. There are already several drones that can hook up to power lines line and recharge. Design Research Associates has developed such a system for small drones and a spin-off known as the "Bat Hook" for Special Forces. This really does look like something Batman would use – a small, sharpened boomerang with a line attached. Toss it over a power line and the sharp edge cuts through insulation. A device on the end of the line converts the high-voltage alternating supply into a regulated direct current for charging electronics. The Bat Hook can be popped off the line so it can be removed and reused.

The Urban Beat Cop version of the Aurora's Skate has this type of power scavenging built-in, enabling missions that continue indefinitely. The drone can view the neighbourhood from one vantage point, then stealthily fly off and perch on another power line a few blocks or a few miles away.

AetherMachines have developed a third such power-line, power-scavenging drone under a series of Air Force contracts. Their drone weighs less than three pounds, boasts special power-line perching flight control systems, and a universal input that can take power from lines of different voltages. (12)

Being able to perch and recharge is a game changer. Drones are not just tactical devices for patrolling or dealing with a particular incident; they are permanent sentries that can keep watch longer than a Predator. The drone becomes a flying CCTV camera, a permanent part of the urban landscape, capable of carrying out long-term surveillance. The Urban Beat Cop design includes software to carry out some types of pattern-of-life monitoring automatically. It could keep track of the comings and goings of specific vehicles in an area or potentially even individuals.

It may not be Big Brother watching you in the future, but a small perching drone.

While this might promote visions of a future when every drone strike is carried out with perfect intelligence, history suggests it will not be enough to prevent the killing of innocent people.

In July 2005, police in London were staking out a block of flats where bombing suspect Hussain Osman was believed to live. They identified an individual leaving the building as Osman. The police believed he was about to carry out an attack. They pursued him on to a train, shooting the man seven times in the head and killing him instantly. The victim was Jean Charles de Menezes, a completely innocent Brazilian electrician with little actual resemblance to Osman.

While an automated recognition might eliminate some of the errors made by anxious humans involved in a stakeout, mistakes are still possible, if not inevitable, with remote drone observation systems.

Forty years of solar planes

Perching on power lines may work in some places, but military operations often take place in under-developed rural areas; in deserts, tundra, and jungles; as well as far out at sea. One answer is solar power, which has fascinated aeronautical designers for years. Aircraft wings are large, flat surfaces that might easily be covered in solar cells. If a solar aircraft could generate enough electricity, it could power its own engines and keep flying on sunlight alone. If it could generate more electricity, it could charge up batteries to keep it flying through the night as well.

The problem has always been the high weight and low efficiency of solar cells. The first cells in the 1950s converted only about 4% of the energy falling on them into electricity. By the 1970s, this had been increased to a more respectable 11%, and ARPA (the forerunner of DARPA) commissioned Astro Flight Inc. of Irvine, California, to build the first solar aircraft.

The AstroFlight Sunrise was essentially an unmanned glider. Although it had a wingspan of thirty feet, it only weighed twenty-seven pounds. It first took off in 1974, and the four hundred and fifty watts produced by the mass of solar cells was enough to keep it flying for three hours at a time on sunny days. (13)

The next challenge was a manned solar aircraft, and another Californian company, AeroVironment, came to the fore. As we have seen, AeroVironment founder Paul MacCready specialised in lightweight gliding aircraft and had already developed man-powered planes, so he turned his skills to solar-powered ones with the Gossamer Penguin. (14)

The first manned, solar-powered flight was piloted by Marshall MacCready, the designer's thirteen-year-old son – the choice of test pilot being partly determined by the need to be as light as possible. At eighty pounds, Marshall was ideal. The plane was towed by a bicycle to assist take-off, and it only flew five hundred feet, but like the Wright Brothers achievement in 1903, it showed that the thing was possible. The plane, Gossamer Penguin, was so delicate that it could only be flown in the calmest conditions, which typically come just after dawn. Unfortunately, this is a bad time for solar cells, as the sun is at a low angle and little light reaches them. A bigger, tougher plane was called for.

MacCready's next evolution, Solar Challenger, had a wingspan of almost fifty feet and weighed two hundred pounds empty. (15) Pilot Stephen Ptacek added another hundred and twenty pounds. The construction, Kevlar and foamed polystyrene, betrays the family relationship to the Raven. In July 1981, Ptacek flew Solar Challenger the 163 miles from Paris across the Channel to an air base at Manston in England. As is typical for solar planes, Ptacek flew at an average speed of just thirty miles an hour, though he did rise to an altitude of eleven thousand feet before he started the crossing to ensure he would be able to glide to safety if necessary.

Solar Challenger was followed by the even lighter Sunseeker, a manned plane that traversed the continental US by stages in 1990. In the latest development in manned solar aviation, the Swiss-based Solar Impulse team aims to fly an aircraft all the way around the world. Solar Impulse's wings span over two hundred feet and the solar cells built into them produce over forty-five kilowatts of power, a hundred times as much as the early Sunrise. That's still not much by aircraft standards; Solar Impulse has a wingspan six times as great as the popular Cessna 172, but less power.

Solar Impulse is, of course, light for its size at 3,500 pounds, somewhat more than the Cessna. It has a leisurely cruising speed of 43 mph. Its journey is a slow and perilous undertaking, and the plot and mission control team have to keep one eye on the weather because of the risk of storms. (16)

As of late 2015, Solar Impulse's batteries are being repaired following thermal damage incurred during the Japan to Hawaii leg, and the round-the-world trip does not look like it will be resumed until April 2016.

The limitations imposed by solar technology means these manned craft are inevitably fragile, slow machines. They fly more like airships than airplanes, always at the mercy of the winds. The whole enterprise harks back to an age when aviators were adventurous daredevils rather than professionals with customer-service skills. Manned solar aircraft look great, and they might even catch on as an environmentally friendly recreation. But it is hard to see any practical use for them. Most are destined to end in museums because their real goal is to raise the profile of solar power and show its potential, not to push aviation technology forwards. In this they fall into the same category as solar-powered cars as showpieces.

The Pentagon has maintained its interest in solar-powered "eternal aircraft" ever since the AstroFlight Sunrise. AeroVironment quietly built a solar drone for the military called HALSOL using the same technology as Solar Challenger. A long-endurance drone called "High Altitude Powered Platform" was designed under a classified program for strategic reconnaissance. At one point a solar-powered drone was even developed as a component of a Star Wars defense system. This was known as Raptor/Talon combination: a vast solar

Raptor plane would stay permanently aloft just outside enemy territory, watching for ballistic missile launches and intercepting them with tiny Talon missiles. Like most Star Wars hardware, Raptor/Talon remained in the world of fantasy. (17)

The most recent military project was the Zephyr, a craft with a seventy-foot wingspan yet weighing just over a hundred pounds. Launched by hand, it soared to 61,000 feet. In 2010 it flew nonstop for a record-breaking 336 hours – almost two weeks. But the big, fragile aircraft is still seen as impractical. Zephyr was part of DARPA's Vulture project, which has been downgraded from "development project" to "technology demonstrator." In other words, nobody is expecting to see a viable aircraft at the end. (18) The makers believe that a commercial version is possible, but as usual this remains to be seen.

There are others in the commercial world. Titan Aerospace, a company acquired by Google, plans to build large solar-powered drones that will fly continuously at high altitude. These are described as "solar atmospheric satellites," intended to compete with communications satellites. Like manned solar craft, Titan's drones will be large, delicate planes. They will operate in a "sweet spot" in the Earth's atmosphere, an altitude of 60-70,000 feet where the air is relatively calm, the first flight was due in 2015, but a crash in May that year put the schedule back.

The crash is a reminder that of how hard it will be for Titan's drones to avoid the fate of HELIOS, NASA's large solar aircraft, which ran into turbulence high above the Pacific in 2003. HELIOS was torn apart mid-air, and the fragments fluttered down to the ocean, ending NASA's solar aircraft program.

Small Solar

Solar power makes more sense for smaller drones than larger ones. It has to do with the favorable ratio of the wing area – and hence the number of solar cells that can be carried – to weight, compared to larger aircraft. This is the revenge of the scaling factor that puts small drones at a disadvantage when it comes to carrying fuel. A plane that is half the size has one quarter the wing area, and hence only carries a quarter as many solar cells, but it also only has one-eighth the weight to support.

While big drones have to be specially designed to maximize the wing area, making them into vast, flimsy contraptions that disintegrate in a strong gust, small drones have a better ratio of wing area to power requirement. Something on the scale of Solar Impulse generates only a fraction of the power of a normal aircraft of the same size, but the Raven's wings are can generate the modest tens of watts it needs. Even a small drone not optimized for solar power can gain a significant boost, and solar power could give virtually unlimited flight times to a well-designed drone.

A solar recharger is handy for troops out in the field. Rather than needing an external power source, they can just leave the drone in the sun for a few hours. That is quite a bonus if you are on a foot patrol lasting several days. And of course, a perching drone can sit like a cormorant with its wings outstretched, absorbing solar rays.

In 2010, Javier Coba added solar power cells to a Raven as his naval postgraduate thesis project. (19) He used indium gallium diselenide cells that only had an efficiency of 13% but were easily available. The project was highly successful. At a total cost of around a thousand dollars, Coba's solar Raven was flying for 30% to 70% longer than the basic version, even in dull conditions. He calculates that in sunny conditions the same design should more than double flight time.

Coba carried out his project with great expertise but minimal resources. When he finished his thesis, he noted that electronics had become more efficient and it would be possible to build a better version than when he started. More important, the solar cells now available are much better.

Coba's project tells us something important about small drones. He did not need a corporate laboratory or lavish industrial sponsorship, but significantly improved drones working virtually on his own. By contrast, Solar Impulse cost around six million dollars and has little practical application.

While much can be achieved in garage workshops, corporate power is now driving small solar drone technology. Companies with new, high-efficiency lightweight solar cells are looking for applications where they give a real advantage. For rooftop power generation, weight and efficiency are not so important; cost per watt is the key metric. But Rich Kapusta of Alta Devices of Sunnyvale, California, says that once his company developed the thinnest solar cell with the highest single-junction efficiency in the world (20), small drones looked like an obvious market.(21)

Alta Devices' cell is based on gallium arsenide (GaAs) rather than the traditional silicon. GaAs has long been known as a better solar material but has attracted little interest because it is so expensive. Alta Devices' response was to develop techniques for growing very thin GaAs crystals so they require only tiny amounts of the precious material. This was just the start; equally challenging were the new techniques needed to lift off the thin sheets of crystal without damaging them and sandwich them together with layers of copper and plastic. The whole assembly, including its protective encapsulation, is about as thick as a sheet of writing paper. The thin sheets are light, flexible, robust, and comparatively cheap. They have an efficiency of almost 30%, more than twice as high as the cells on early solar planes, and at a fraction of the weight.

A couple of ounces of cells provide all the power a drone like the Raven needs, and the circuitry for charging batteries from the solar cells adds only a few ounces more.

These GaAs cells are produced as five-by-two centimeter units that can be stitched together to cover a wing of any size and shape. As a demonstration, AeroVironment and Alta Devices worked together on a special solar version of AeroVironment's Puma UAS. Fitted with Alta Devices lightweight solar cells, the solar Puma flew for over nine hours, three times as long as the pure battery version.

In late 2014, Alta Devices joined with Airware, a company that produces plug-and-play add-ons for small drones, in a partnership aiming to make solar power accessible for drone makers.

Meanwhile, Microlink Devices of Niles, Illinois, has developed their own solar sheeting based on a different solar technology, a GaAs triple junction (22). It is tuned to pick up light across three separate wavelengths rather than one; the total efficiency is similar at around 30%. The layer of photovoltaic material mounted on a sheet of metal looks like thin metal foil. Although brittle, it is flexible enough to be wrapped around a cylinder just a few inches in diameter and can conform to the curves of a small wing.

Microlink has worked with the US Air Force Research Laboratory to retrofit solar cells to AeroVironment's small drones including Raven. Their first solar-assisted Raven showed a 60% improvement in endurance, and this should increase to 100% with further work. Other drones with a larger wing area, like Aurora Flight Sciences' Skate, would do better.

This sort of improvement is dwarfed by the capability of a small, custom-built UAS which Microlink is collaborating on with Design Intelligence Incorporated (makers of the perching drone mentioned earlier). It is called Eturnas.

"By treating solar power as an 'add on' or addition to an existing UAV design, we might be able to achieve 200% to 300% improvement in flight endurances over battery power alone," says James Grimsley, President and CEO of DII. (23) This figure is certainly borne out by the solar Puma. "However, for a solar optimized vehicle such as Eturnas, we can literally fly as long as there is good sunlight. Since Eturnas also has battery power, it can easily operate for another hour or two after the sun goes down."

The first small solar drone is already on the market. The Solar Falcon, a hand-launched drone for security and other applications made by a company with the same name, uses thin-film cells for a claimed endurance in excess of seven hours in good conditions.

Drones can even harvest some energy after dark. Aurora Flight Sciences has tested a version of their Skate with solar cells on the upper wing surface and infrared cells on the underside. These only provide a fraction of the power of solar cells, but as the thin film weighs virtually nothing they are a bonus when a drone already has the necessary recharging circuits.

Solar power may also benefit from the smartphone industry. Some phone makers, including Apple, are interested in built-in solar charging. The quality of small-scale solar cells is likely to improve, and large-scale manufacturing facilities are likely to send prices plummeting.

Small drones have a disadvantage when it comes to storing power, as noted above. Big solar aircraft like those envisaged by DARPA's Vulture project and Titan Aerospace's satellite-like communication platforms can store enough energy to last the night when the sun goes down. At dawn, they switch to solar power and start recharging their batteries, and so keep flying indefinitely. Small drones are again stuck with the limitation of only flying as long as the battery lasts.

However, the gap is closing fast. The Swiss team AtlantikSolar aims to send a hand-launched fifteen pound solar drone across the

Atlantic. In June 2015 they achieved a flight in excess of twenty-eight hours. The flight ended with the drone's batteries fully recharged. This is a huge leap forward, showing that under the right conditions, a small solar drone can fly indefinitely with existing technology. (24)

Solar power may also be combined with perching. Like birds, flocks of drones might fly by day and roost at night, conserving power until the sun comes up to recharge them. The flock would still keep some of its members in the air continuously, even though no individual has enough charge to keep flying all night. They could maintain continuous coverage, just as the Predators that make up a CAP constantly take over from each other.

A drone that combines perching with recharging from power lines and the sun can operate for extended periods. But there is another approach that does not require any extra hardware, and birds have been doing it for millions of years.

Soaring Forever

The air itself can keep a flier aloft without expending any power. Soaring birds have perfected the art of riding air currents: eagles ascend on thermals, albatrosses cross oceans with the aid of wind shear, and even the urban pigeon gets a boost from wind gusts. Migrating birds fly for days without touching down, taking advantage of air currents even in their sleep.

This is another situation where small drones are in a favorable situation; you could not hope to support an airliner on the wind alone, and manned sailplanes require outsize soaring wings. But drones the size of birds can soar like an eagle.

Kevin Jones, a research associate professor at US Naval Postgraduate School in Monterey, California, is part of a team developing a hand-launched drone called TALEUAS (Tactical Long Endurance Unmanned Aerial System) which combines solar power collection with an ability to soar on thermals. (25). it is based on the airframe of a radio-control model used in cross-country gliding competitions.

The wings of TALEUAS are covered with monocrystalline silicon solar cells, chosen for their low cost and easy availability. These have given the team plenty of engineering challenges, as the cells are brittle and more prone to crack than the solar films produced by Alta Devices and Microlink. But his work is equally focused on soaring.

TALEUAS is designed to soar upwards on thermals like vultures and eagles. Thermal updrafts are bubbles of warm air that form where the sun heats the ground. Because warm air is lighter, the bubble tends to separate from the ground and rise upwards. Soaring birds, which would require a huge muscular effort to climb by flapping, spread their wings and ride these thermals upwards. TALEUAS can do exactly the same. It is launched with the aid of its electric motor, but the team has developed a technique for it to find and lock-on to thermals, maneuvering continuously to stay inside one when it finds it, and rising thousands of feet.

Even on mild days with temperatures in the 50s, TALEUAS can climb to over four thousand feet. The drone is actually better at soaring than some birds. Jones says that raptors that came over and tried to share his thermal found it was not strong enough to support them. (26)

Like soaring birds, TALEUAS can ride thermals to climb and stay aloft for long periods without expending any effort. Soaring birds stay up all day with this method. They can also use it to travel across country, gaining altitude to glide long distances before finding the next thermal to lift them up again, and so continuing onwards.

Birds are astute when it comes to finding thermals. They can recognize the right sort of landscape; car parks tend to be good as the black surface heats up quickly. They can spot floating dust clouds and bits of vegetation that show how the air moves. When Paul MacCready was testing his Solar Challenger in the Arizona desert, he found that a gentle thermal was a great help; during test flights he had a van drive around raising clouds of dust that helped highlight where the thermals were. In more humid areas, fluffy cumulus clouds may form where the updraft carries moisture to a high enough altitude for it to condense.

Birds are better than glider pilots when it comes to finding a thermal. This led British falconer Scott Mason to invent a new sport: parahawking – paragliding with a bird as a guide. The birds – kites or vultures – are trained to lead a paraglider pilot to updrafts. (27)

Based in Pokhara, Nepal, Mason offers travellers a parahawking experience on a tandem paraglider. The bird returns to the paraglider at intervals to claim a food reward for each thermal it finds. Mason developed parahawking as an "enrichment exercise" for his rescued birds, providing them with more entertainment and exercise than being flown from a falconer's wrist on the ground. But parahawking also shows the advantage of having a guide when looking for thermals.

Birds sometimes locate thermals by watching each other; when one vulture finds an updraft, the others will glide over and take advantage of the discovery. (They do not always share with strangers though. Jones has seen a falcon attack a bald eagle, and on occasion a hawk dived on TALEUAS and try to drive it away from a thermal.) Like a flock of vultures, Jones suggests that drones could work cooperatively. Between them, they could cover a wider area. As soon as one finds a thermal, the others can come and share it. Between them, a group of drones could quickly map out the landscape in an operational area and find where the thermals are likely to form at different times of day.

TALEUAS' combination of solar cells and soaring allow it to stay in the air for as long as there is sunshine or an updraft. Flying all night would be a matter of battery power, but again with a combination of perching and solar power, a swarm would be able to keep at least one drone in the air to maintain eyes on the target 24/7.

Future developments will include more advanced ways of finding updrafts. Some researchers have already experimented with using thermal imagers and other sensors to spot them from a distance. Mike Hook and his team at Roke Manor Systems studied the geography of updrafts. This includes not only thermals but also orographic lift, produced when wind blows over a ridge, and lee waves created where wind strikes the land.

The resulting software combines landscape analysis with cameras to spot the telltale cumulus clouds and replicates the skills of an experienced sailplane pilot – or a vulture – in knowing where to find rising air. Hook flew his prototype in a sailplane, and the pilot was impressed that the system was able to spot thermals he might have missed. Even this early prototype suggests that with the right software, any drone with a camera can soar and fly for longer.

Long-Distance, High-Speed Robot Albatross

Phil Richardson is an oceanographer at the Woods Hole Oceanographic Institution in Woods Hole, Massachusetts. He became fascinated by the idea of dynamic soaring after watching albatrosses over the Southern Ocean. He was in a research ship, and he watched the birds effortlessly keep pace without ever flapping their wings. This was in spite of the fact that the ship was steaming ahead at ten knots, and the birds were flying against the wind. (28)

Albatrosses exploit wind shear, the difference in wind speed at different altitudes. There is little wind close to the sea, but if an albatross gains a little height, it catches the wind and is carried upwards like a kite in a breeze. The bird then turns and glides down in the direction it wishes to travel, converting its altitude into speed. When it starts to slow down, the albatross angles its wings to rise and catch the wind to repeat the maneuver. Albatrosses can soar even in their sleep, flying thousands of miles without effort. One snowy albatross was tracked flying four thousand miles in twelve days.

After writing a paper on the mathematics involved, Richardson watched radio-controlled glider enthusiasts in Southern California harnessing an extreme version of the effect, using orographic lift, wind from the sea that was diverted upwards by cliffs. By repeatedly looping over the cliffs, the little gliders can build up tremendous speed.

"I was totally blown away by how fast they were flying – up to five hundred miles an hour," says Richardson.

Flying a model glider is a fine art, and the enthusiasts know exactly where to go to find an updraft. But dynamic soaring using wind shear is an even more subtle technique and, until recently, there was considerable debate about exactly how it works. It was only in 2013 that Gottfried Sachs and his colleagues at the Institute of Flight System Dynamics at Munich Technical University finally proved that the albatross does indeed use the exact technique described above. Using extremely precise miniature GPS trackers on birds, they showed that the flight pattern has four stages: a windward climb, an upper curve, a leeward descent, and a lower curve, all repeated over and over. Exactly the predicted pattern for dynamic soaring.

Experimental flights back up the theory. In 2006 NASA carried out test flights over the dry lake bed at Edwards AFB with an L-23 Super Blanik sailplane. The aircraft dipped in and out of the boundary-layer wind shear in a number of different flight patterns and achieved a significant boost from the dynamic soaring effect. However, because the plane was so large and heavy, it could not get nearly as much benefit from dynamic soaring as an albatross.

In 2015, Richardson produced a new paper on how an albatross or a drone can pull off the trick he witnessed in the South Atlantic of gliding against the wind. He calculates that a bird can fly at a forty-five degree angle into the prevailing wind. By zigzagging from left to right of the wind, it can achieve a net forward motion. The effective upwind speed is about 60% of its actual speed. This matches Richardson's observations, when he noted a bird keeping pace with the ship made two turns over each ten-second period. (29)

Richardson has looked at how this approach could be applied to an ocean-surveillance drone based on the 100 DP, a radio-controlled hobby glider. The 100 DP can handle greater forces and exploit faster winds than a gull. It can fly in any direction, regardless of the wind, by tacking like a sailing ship and taking a zigzag course. The theory suggests it could fly upwind at over five times the speed of the wind. Flying diagonally, the fastest mode, it could make over eight times wind speed. The robot albatross could carry out extended surveys or searches at 100 mph or more, without using a drop of fuel. Such a drone could cross oceans as easily as the albatross.

This model may need some adjustments. As Richardson notes, the albatross can fly so low its wing tips dip into the water, but an aircraft trying the same maneuver risks crashing and will need bigger safety margins. But the principle is sound.

The challenge is making it work in practice – a challenge that has been taken up by Jack Langelaan and colleagues at Penn State University. They are developing an autopilot to carry out dynamic soaring, with the first flight scheduled for late 2015. Langelaan's team has been researching the area for three years, and they have broken it down into three key areas: wind field estimation, path planning, and flight control. (30)

"Wind field estimation is about determining where the shear [difference in speed at different heights] is and what it looks like," says Langelaan. "The data from this is passed to the trajectory planner, which computes a trajectory which is at least energy neutral but ideally gains energy from the wind shear. The flight controller turns this plan into action."

The drone learns about where the airflow is as it flies, so its model of the wind field is continually updated. The team has carried out numerous computer simulations and hardware-in-the-loop lab tests to put the trajectory planning software and flight controllers through their paces. It is like a flight simulator for electronic pilots.

The test plane is a commercial radio-controlled glider weighing thirteen pounds with a wingspan of nine feet. It is not optimized for dynamic soaring; the initial aim is simply to show that automated dynamic soaring is possible. The first flights will involve the maneuver performed by Californian glider enthusiasts witnessed by Richardson: flying closed loops above the ridgeline to exploit the airflow where wind comes over a ridge. The glider enthusiasts aim for maximum speed by going faster and faster with each loop, but Langelaan just wants a robust, working system. His aim is endurance: if it works, it will show that a soaring drone could, like an albatross, keep flying for days or weeks.

Dynamic soaring should also work on land in some places. Sachs' team has showed that in addition to thermals and ridge lift, some soaring birds use dynamic soaring over land. Deserts and plains would be the most promising, but wherever there is a difference in wind speed with height, dynamic soaring should be possible.

Urban fliers

Perhaps the toughest places for small drones are the urban canyons of big cities. Tall buildings create their own patterns of wind circulation. This was a well-known feature of the Flatiron Building, the first skyscraper completed in Manhattan in 1902, which channeled the wind around it. Idlers hung around the base of the building so they could get a glimpse of women's ankles as the wind blew up their long skirts.

This is a dangerous environment to fly a small aircraft, but Reece Clothier and colleagues at the Royal Melbourne Institute of Technology (RMIT) in Australia are developing a drone that can take advantage of the wind around tall buildings. The inspiration came when Clothier watched birds riding the updrafts between skyscrapers in Melbourne and wondered if it would be possible for a drone to do the same. The Kestrel project work combines computer modelling of airflow around buildings with special wind sensors for small drones. The idea is that the drone will – in Clothier's words – "surf the updrafts" associated with tall buildings. (31)

Just as Langelaan's software provides an automated awareness of wind shear and how best to use it, Clothier's Kestrel will have a built-in understanding of urban wind patterns. By taking advantage of them it will also able to fly like a bird, gaining a significant increase in endurance and avoiding potentially damaging turbulence. The project is being sponsored by Australia's Defence Science Institute.

Like the TALEUAS though, flying like a bird can lead to Kestrel being mistaken for one by other birds. One prototype was lost when an eagle dived on it and carried it off.

As well as the treacherous winds of urban canyons, there are more general winds, breezes, and sudden gusts that can catch a plane. Airline passengers are well aware of "clear air turbulence," which is simply unpredictable wind at high altitude. Running into turbulence at high speed can damage an aircraft, or throw passengers about in the cabin, which is why it is always a good idea to keep your seatbelt buckled.

These random winds can be a handy source of energy. The REactive Controller for HArvesting Gust Energy (RECHARGE) is being developed by Area-I Inc. of Canton, Georgia, for the US Air Force. It comprises sensors and software to detect local airflow and automatically steer into the wind. An airliner is too big and fragile, and lacks the agility for such a maneuver; birds do it without thinking. There is no reason why small drones cannot do the same.

Turning into a gust of wind produces lift, translating the wind into extra altitude and increased endurance for no fuel expenditure. RECHARGE makes the aircraft more stable as well as reducing sudden stresses on the airframe. Turbulence puts strain on the wings and buffets the aircraft; minimizing it allows the drone to ride out gusty weather that can ground other aircraft.

The developers say that when RECHARGE technology is mature, it could provide a simple upgrade and give "significant performance benefits in terms of range and endurance" to thousands of small drones like Raven.

RECHARGE relies on gust detection from the effect on flight, but Aurora Flight Sciences is working on an exotic sensor which will provide faster reaction. Birds and insects use proprioception, direct feedback via feathers or hairs or the forces on their wings to respond instantly to changes in airflow around them. Aurora are exploring sensors embedded in a small drone's skin to give it the same sensing capability, so it could detect sudden gusts more quickly, taking appropriate action via the sort of flight control software being developed by RECHARGE. (32)

Rather than tearing an unmanned aircraft apart, like NASA's unfortunate HELIOS, a windy day might give future drones a lift. Soaring technology is still in its infancy, but I have talked to researchers who would like to harness wind power directly to recharge their drones simply by using the propeller as a wind turbine. This will require extra circuitry, so there is a weight penalty, and some clever flight software is needed to cope with the extra drag. This is challenging when soaring has yet to be fully mastered. But in a good breeze, future small drones may recharge themselves entirely from wind power.

We know from nature how well air currents can be exploited. As anyone who has watched gulls or other soaring birds will know, they have an enviable agility in strong winds, and can wheel, dive, and gain altitude at will. Perching birds will wait for a breeze and take off vertically, then swoop down, using the altitude they have gained to fly upwind at speed. The wind that currently keeps small drones grounded enables birds like this to fly so well. And since it is largely a matter of understanding how air currents move and how they can be exploited, this expertise can be turned into software and downloaded on to drones.

Flying into the future

We have looked at various techniques for beating the limitations of lithium batteries: better battery chemistry, perching, scavenging energy from power lines, solar power, and ways of getting a lift from air currents. All show promise, and none of them are mutually exclusive. There are already designs for perching drones that have solar cells as well as power scavenging. TALEUAS integrates soaring and solar power into the same aircraft.

It is not fanciful to believe that small drones will be able to carry out missions lasting weeks. In fact, the Air Force's Micro Munition program involves a small drone that can do exactly that; some of the projects mentioned above have been funded as part of this effort. However, the Micro Munitions drone is designed for a one-way trip only, perching or loitering around the target area until it gathers information and receives the order to strike.

In time, one or more of these technologies is likely to dominate. A craft with more efficient solar cells and better batteries would not need any other power source. Batteries are likely to improve significantly, and the 30% efficiency of solar cells will be superseded. Devices with an efficiency of 37% and more already exist in the laboratory and will be commercialized in due course.

Alternatively, stealthy perching might turn out to be so tactically useful that other approaches will be redundant. Because they can be acquired in such numbers, small drones might be more like the unattended grounds sensors already in use (which we will see in Chapter 7): so many can be dropped from an aircraft that they hardly need to move around, but can lurk in place until they need to attack.

Other, more exotic technologies may also develop rapidly. Researchers have demonstrated robots that can "digest" leaves or other material from the environment. The power levels are currently tiny, but clearly this energy source works well for birds. Beamed power, provided by a laser or radar beam, has long been used to recharge small drones from long range in demonstrations. Some companies are working on turning this into a practical technique rather than a laboratory gimmick. Such technologies are generally no use to bigger aircraft, but benefit small drones. It is another of their advantages that they can be easily redesigned to take advantage of a new component and do not have the twenty-tear generation time of manned combat aircraft.

Whatever direction the technology takes, future small drones will have far longer mission times and ranges than the small drones of today. A swarm of small drones released from a transport aircraft mothership could blanket an area and loiter in place for as long as they are needed. They might also transport themselves, "self-deploying" as the Air Force calls it. Small drones could fly intercontinental or transoceanic distances like flocks of migrating birds – or like the Japanese Fu-Go fire balloons.

Given this sort of capability, small drones are in a position to challenge the Predator and its relations for a place in the larger strategic arena.

The increase in small drone endurance from hours to days is astonishing. But as technology progresses, small drones will not simply be able to operate for longer. Advancing technologies will make drones vastly more capable than those we have seen so far.

REFERENCES

1) TAM-5 Transatlantic flight

Barnard Microsystems. "Long distance record for FAI Class F8 model plane." (http://bit.ly/1LRdqTf)

2) Tesla Model S Battery

Green car congress: EPA rating for 85 KWh Tesla Model S: 89 MPGe, 265-mile Range. (2012, June 21). (http://bit.ly/1DCgxAb)

3) IBM Lithium Air Battery

Lithium/Air Battery Project (Battery 500). (n.d). (http://ibm.co/21yD9LI)

4) Molten Air Battery

Zyga, L. (2013, September). Molten-air battery's storage capacity among the highest of any battery type. (http://bit.ly/1QhQHWB)

5) OXIS Battery

A whiff of brimstone. (2014, December 30). (http://econ.st/1OAaNLG)

6) Stalker XE with Fuel Cell

Hambling, D. (2013, May 3). Longer-lasting UAVs powered by Fuel Cells. (http://bit.ly/1QWCJer)

7) Ion Tiger fuel cell drone (n.d.).

8) Urban Beat Cop

New 'urban beat cop' surveillance system to keep military operators out of harm's way. (2014, August 14). (http://1.usa.gov/1TnvkCH)

9) Vishwa Robotics (http://bit.ly/1LRdrqi)

10) "If you could perch you could look around without using any power." – Bhargav Gajjar interview with author December 2013

11) ImageNAV-LZ: Autonomous Landing Zone Detection. (2013, December 9). (http://1.usa.gov/1LRdxy4)

12) AetherMachines perching, power scavenging drone (n.d.). October 24, 2015 (http://1.usa.gov/1NsDOIp)

13) Astroflight Sunrise

Boucher, R.J. (1985). Sunrise, the world's first solar-powered airplane. J. Aircraft, 22(10), 840-846. (http://bit.ly/1TnxTEG)

14) Gossamer Penguin

Russ, L. (2010, August 7). First Public Demonstration of Solar-Powered Gossamer Penguin. *Smithsonian National Air and Space Museum.* [Web log post]. (http://s.si.edu/1NJIXxN)

15) Solar Challenger (n.d.). (http://bit.ly/1IGoH8T)

16) Solar Impulse RTW. (n.d.). (www.solarimpulse.com)

17) Revive Raptor Talon! - High Frontier. (2014, May 29). (http://bit.ly/1jB743E)

18) Zephyr UAV continues to break records on first authorized civil flight. (2014, September 28). (http://bit.ly/1QWFeNP)

19) Solar Raven project

Coba, J. & William, V. (n.d.).. *Application of Copper Indium Gallium Diselenide photovoltaic cells to extend the endurance and capabilities of the Raven RQ-11B Unmanned Aerial Vehicle.* (Naval Postgraduate School thesis). (http://bit.ly/1NsaGN8)

20) Alta Devices

Lightweight, flexible solar cells increase UAV flight endurance. (n.d.). (http://bit.ly/1Q3eU4z)

21) "Small drones looked like an obvious market" – Rich Kapusta interview with the author, August 2013

22) Microlink Devices

Microlink devices produce page. "Photovoltaics" (http://bit.ly/21yH6Qs)

23) "By treating solar power as an 'add on.'" – James Grimsley interview with author, August 2013

24) Atlantiksolar (http://bit.ly/1NJJq33)

25) TALEUAS

Flights of fancy. (2013, February 8). (http://econ.st/1PABFNi)

26) "Raptors which came over and tried to share his thermal found it was not strong enough to support them." – Kevin Jones interview with author, January 2013

27) Parahawking

The parahawking project - The official parahawking website. (n.d.). (http://bit.ly/U6M48v)

28) Dynamic Soaring

Hambling, D. (2013, July 3). The future of flight: Planes that never need to land. *Popular Science*. (http://bit.ly/1NsbCRC)

29) Phil Richardson robot albatross

Hambling, D. (2014, December 30). Wind riders: Amazing albatross flight inspires drones. *New Scientist*. (http://bit.ly/1TAY9MF)

30) Jack Langelaan Penn State University project

Jack Langelaan Interview with author, December 2014

31) Kestrel

New project to develop autonomous soaring aircraft. (2014, April 7). (http://bit.ly/1QhVoj6)

32) RECHARGE (http://bit.ly/1lyRAOM)

Reactive controller for harvesting gust energy. (n.d.).

CHAPTER SIX

FASTER FORWARD

> *"Wait a minute, wait a minute I tell you – you ain't heard nothin' yet!"*
>
> - Al Jolson, The Jazz Singer

In the ten years from 2003, the Raven went from an obscure Special Forces gadget to the most common military drone on the planet. Over ten thousand have been produced. As we saw in Chapter 3, the flight time, sensors, and software of the Raven have steadily improved. Unusually for the military, there were no massive price rises. Unlike the Predator/Reaper, which is evolving slowly and seemingly into a dead end, the small drones have developed more like consumer electronics. The next decade promises even faster evolution. Three forces contribute to this acceleration.

Firstly, there is now an existing user base. Ten years ago, there were only a handful of small drone operators; there was no market to sell to, and no group of users looking for enhancements. Now there are thousands of military users, and the civilian drone business is taking off.

While they remain in legal limbo as far as the FAA is concerned – it is illegal to use drones for commercial purposes in US airspace without special licensing – the number of drone suppliers and would-be users is increasing fast. Thousands are already used in the US for scientific research, farming, filming, and a growing range of other tasks. Build a better drone or drone component, and there are plenty of people to buy it or who will supply funding for development.

Secondly, the seemingly unstoppable growth of the electronics industry in general and the smartphone sector in particular is driving advances directly applicable to small drones. Every generation of smartphones is more capable than the last, and this means cheap electronics available for drones.

Thirdly, a force even more powerful than the electronics industry: harnessing the power of nature. Scientists are increasingly developing biomimetic technology, which borrows techniques from living things. Some of this was evident in the previous chapter. Jones' TALEUAS soaring like an eagle, Clothier's urban-surfing Kestrel, Richardson's robot albatross, and Gajjar's perching legs are all biomimetic. Evolution has spent millions of years optimizing its creations, and researchers are leveraging these readymade "perfect solutions" in areas including vision and flight control.

As with smartphone technology, biomimetics benefits small drones more than bigger aircraft. Airliners have little in common with swallows or bats, whereas small drones are basically artificial birds. An RQ-11 Raven may not be a match for a feathered raven in terms of agility or intelligence, but future drones will increasingly challenge the capabilities of their avian counterparts in all sorts of ways.

An additional enabling factor is that the cost of entry is so low that development is now in the hands of students, one-man companies, and garage inventors. As we saw with MITRE's Razor project and the miniature Crazyflie quadcopter, open source is an effective model for small drone development. Garage inventors cannot build their own Predator, but they certainly can build their own Razor, and any good ideas can be picked up and adopted by the community of small drone makers.

Chinese drone company DJI is actively encouraging this with the introduction of the Matrice 100, which they call "a quadcopter for developers," a testbed to try out new hardware and new software. (1)

"Develop a system for any use, and revolutionize industries by applying your knowledge and skills to cutting-edge flight technology," DJI suggests on their product page.

With various expansion ports and a design to make modification easy, the M100 should set off a new round of drone development, one based on DJI's hardware. Whereas once manufacturers would have discouraged anyone from hacking their hardware to build "mods" or upgrades, DJI anticipates this will work in their favor.

Even without this sort of assistance, small-scale developers have come into their own in the drone world. Kickstarter has successfully launched a number of small drone projects, and more are in the pipeline. In the time it takes a major aerospace company to agree on the budget levels, staffing, and scope of a new project, a start-up company can get a product to market and be on their way to version 2.0.

Upgrades

The Raven upgrade spiral was driven by the pull from users and the push from newly-available technology. Both continue. Like most military technology, the Raven's original control system was a comparatively clunky affair compared to the slick user interfaces on smartphones. Considering the millions of users and extensive development that goes into consumer devices, this is hardly surprising. Now that Raven has an established user base, upgrades are on the way.

The Navy's Space and Naval Warfare Systems Center Pacific has developed an improved small drone controller with an intuitive, easy-to-use interface, more like a consumer device. It replaces the tablet-like Ground Control Station and laptop with a single unit. The new controller allows the operator to control several drones at the same time, useful when cheap drones are fielded in large numbers.

DARPA has pushed the multi-drone idea further with a project called HURT – "Heterogeneous Urban RSTA [Reconnaissance, Surveillance, and Target Acquisition] Team." The operator does not control an individual drone; they simply highlight the object or area they wish to look at. A drone is automatically assigned to the task, and the video appears. The drones are autonomous and there is no piloting. This is a step towards the swarming approach we will examine in Chapter 7.

The Raven is currently limited to line-of-sight operation, though this can be extended by relays. Satellite communications, like those on the Predator, have been too large. As the usual miniaturization process gets under way, high-frequency X-band satellite communications are now possible with more compact hardware. Hughes Network Systems is working on a satcomms package suitable for small tactical drones.

While AeroVironment has clearly made a good job of the Raven's airframe, some users may want different versions for particular missions. MAV6 LLC, a company founded by individuals with experience modifying Ravens in the field, has patented their own upgrades. One development is a bulbous but aerodynamic hollow wing with a cavity for a payload. The payload may be sensors, antennas, additional batteries, or an explosive warhead. (2)

According to MAV6, the new wing adds an extra pound of carrying capacity, effectively doubling the Raven's load, without affecting its flight endurance. This is the same sort of modular thinking evident in the Razor with its alternate sets of wings for long-distance soaring or high speed.

The Predator carries out electronic warfare missions, identifying and locating radios, from ten thousand feet. This sort of task previously required equipment too big for a small drone, but Moore's Law is at work. SARA Inc., a leader in military electronics, has developed a range of payloads for intercepting radio signals in different bandwidths, from VHF to the GSM-1800 range used for cell phone traffic. They can locate any radio emitters within range and send back details to the operator's map display. The drone can go and take a look at the radio user on video or simply track the emitter's movements. (3)

In counterinsurgency operations, radio locators can find adversaries using walkie-talkies or mobile phones. Against more capable opponents it will locate jammers, radar, and anyone sending radio signals to their own drones. It could also locate friendly radio users, a process known as Blue Force Tracking, to prevent friendly-fire incidents, or find radio distress beacons during search and rescue operations.

More advanced electronic warfare capabilities are not difficult to imagine; in fact, we already know what they could look like. At the 2011 DEFCON hacker conference, Mike Tassey and Richard Perkins showed off their own Wireless Aerial Surveillance Platform, or WASP. This is based on a drone the size of a Raven, and aims to show just what mobile attackers can do against existing infrastructure.

WASP can mimic a cell phone tower so that phones connect to the network through it, allowing the operators to listen to any traffic. It can also detect Wi-Fi networks and has software to hack into the networks automatically. WASP can be sent on a preplanned flight path to locate a given computer network, break into it, and download targeted data. It is easier than having to break into an installation in person, Mission Impossible–style, and harder to trace. WASP will not leave fingerprints or crack under interrogation. (4)

WASP was developed by people who are essentially amateurs; with bigger budgets and more developers it may be possible to do more. A company called Azure Summit has already developed a military SIGINT (signals intelligence) package for small drones. Its capabilities are not known, but they are likely to at least rival those of WASP.

Combine this sort of electronic eavesdropping with a perching capability and you have a spy who can go anywhere covertly, leaving no trace of its activities. The drone might deliver a payload of viruses or other malware to disrupt enemy systems, adding a whole new dimension to the hacker threat.

Designated Targets

The Raven's most conspicuous battlefield shortcoming is the lack of a laser designator to highlight targets for Hellfire missiles and other guided munitions. It has a laser illuminator so it can "sparkle" targets, highlighting them for helicopter gunners or others to aim at, but it cannot "lase" a target to guide a missile to the aim point. Designators are becoming increasingly valuable on the battlefield as the number of laser-guided munitions has climbed steeply.

Laser designators are large because they have to put out a beam powerful enough for the laser seeker on an incoming missile to lock on to. Predictably enough though, seekers are getting better and the power needed for a seeker to "see" the laser dot has dropped by a factor of four since the 80s. Also, lasers are getting more compact.

Back in the early 2000s, the US Army's "portable" laser designator weighed almost forty pounds. The latest version is below ten pounds. As recently as 2009 there was no designator small enough for the RQ-7 Shadow, a drone with a fourteen-foot wingspan. Now designators are shrinking further, to the point where they can be mounted on Raven-sized drones.

Making lasers smaller is technically challenging, but as with most electronics, it is simply a matter of time. Californian tech firm Areté Associates has developed an "ultra-compact" laser designator module called AIRTRAC about the size of a golf ball weighing less than six ounces. This was built under a US Air Force contract and is just the right size for small drones, mentioned by the contract as a likely application. (5)

Meanwhile Israeli company Elop is already advertising a miniature designator in its Rattler range that weighs just four ounces and was also designed with small drones in mind. (6)

Previously laser guidance was confined to bigger and more expensive weapons like the Hellfire and 500-pound bombs. These days, there are small, cheaper laser seeker heads. Not only are there laser-guided artillery rounds and rockets a quarter the size of Hellfire, but even mortars can fire laser-guided bombs. The mortar might have a range of ten miles, but still needs someone within eyeball range of the target to light it up with the designator. If artillery spotters could do this with a small drone, they would be able to direct precise fire to anywhere.

The small designator opens up a range of new roles for the Raven. In addition to acting as point-man for artillery, there is already some work using them as "off-board sensors" for gunship aircraft and helicopters. The drone is released from an aircraft and flies in to the danger zone to give the pilot or gunner a close-up view. They are particularly valuable because, unlike large drones, they operate at low altitude of just a few hundred feet which puts them beneath the cloud cover. When the cloud ceiling is as low as a thousand feet, Predators and the like are limited in their ability to lase targets.

Without the need to get up close, pilots can release bombs and missiles from long range. Standard, unpowered laser-guided bombs can hit targets fifteen miles away if the plane dropping them has enough speed and height. More advanced gliding bombs, like the recent SDB II, have a range of up to forty miles and almost qualify as drones in their own right.

Air strikes can be carried out by simple "bomb trucks" like transport aircraft which have no need for high speed and maneuverability to avoid anti-aircraft defenses. The enemy would only ever see expendable unmanned drones, loitering overhead permanently and holding everything underneath at risk of instant, laser-guided destruction.

Laser Eyes

As far as sensors go, small drones are limited to cameras, thermal imagers, and the sort of electronic warfare devices described above. Now shrinking electronics are making other options available. Laser-based radar, known as LADAR or LIDAR, works by bouncing a laser off thousands of points. It calculates the distance to each point and builds up a three-dimensional map of its surroundings.

LADAR is more useful than a camera for robots because the result can be used directly. LADAR can measure the exact size and position of objects, whereas with cameras (even stereo cameras), working out depth and scale is challenging. Distinguishing a pothole from a manhole cover, or a tree-shadow from a fallen branch, can be difficult for video-based systems, but it is easy with LADAR. Robotic vehicles like Googles' Self-Driving Car almost invariably have LADAR sensors for this reason.

LADAR also works just as well in pitch darkness, through fog, snow, dust, or smoke as it does in broad daylight. This makes it particularly valuable inside buildings and underground. A robot with LADAR can build up a three-dimensional map with inch-perfect accuracy of a building or cave interior as it goes.

As you would expect, LADAR is shrinking. Previously, such devices weighed several pounds. Velodyne, a Californian company that started developing LADAR as part of DARPA's Grand Challenge for driverless cars, has announced a new system for small drones weighing slightly more than a pound. This gives a three-hundred-and-sixty degree view of the world, mapping out three hundred thousand points a second out to over a hundred yards. The initial cost will be around $8,000, a fraction of the price of earlier systems.(7)

Much of the weight, cost, and complexity of LADAR is due to the precision machinery needed to sweep a laser beam across the scene. Vescent Photonics of Denver, Colorado, has developed technology for miniature LADAR with no moving parts, known as SEEOR for Steerable Electro-Evanescent Optical Refractor. (8) The laser beam goes through a glass waveguide with a special liquid crystal cladding. This liquid crystal has an unusual property: applying an electric voltage across it changes its refractive index. When you change the voltage, the angle of the beam emerging from the glass changes. (9)

Scott Davis, cofounder and Vice President of Technology, says this SEEOR has allowed his team to develop a new solid-state LADAR without moving parts. The initial prototype is the size of a paperback book, but Davis expects this to fall rapidly though the next generations, with progressive versions being the size of a deck of cards and then matchbook size.

The miniature SEEOR LADAR will have a range of a few hundred meters and will be a useful sensor for self-driving cars and other devices. When in mass production, it will be as cheap as other microelectronics, tens or hundreds of dollars, giving small drones an important new mapping and sensing capability.

Seeing through Walls

Radar has always been too big to fit on small drones, but predictably enough, it is also shrinking to fit. Much of this is due to the consumer side of the market, specifically the trend for radar in top-end automobiles.

Researchers at the University of Denver have developed miniature radar to allow small drones to sense and avoid other aircraft. Like LADAR, radar is useful because it works in all weather and at night. The prototype currently weighs twelve ounces, and it is being commercialized by Integrated Robotics Imaging Systems in Kenai, Alaska. The CEO, Joe Parker, estimates it will get down to four ounces. (10)

Some researchers are already suggesting that radar could be built into miniature drones. The Air Force Research Laboratory has funded work on a "smartphone radar" based on the idea that phones already incorporate the three essential components of a radar system: a radio transmitter, a receiver, and computing power. One smartphone on its own is limited, but researchers have looked at using three smartphones together to effectively create a single large antenna.

Most forms of radar are blocked by walls and other solid objects, but there are also through-the-wall "life detection" radars. These use ultra-wideband radio waves, which can go through solid walls as easily as air. Life detection radar works on the Doppler principle and only detects movement, in particular the movements associated with human life – breathing and heartbeats. First responders use portable units in rescue operations to detect if there are people alive in a building. It is imprecise but will give the number and approximate location of any people inside. The technology is also used by the military and police.

In 2013 Israeli company Camero demonstrated a radar system mounted on a small quadrotor – "Airborne structure penetrating life detection system" – which can land on a building and scan it for occupants. Again, this is less of a magical x-ray device that reveals everything, more of a rough indication of whether anyone is home.(11) A similar device for small quadcopters was unveiled by Intellinet Sensors of California in October 2015.

Another approach to through-the-wall sensing is to use existing signals that go through walls. These passive sensors do not require a transmitter of their own but sense objects via distortion in the signal from Wi-Fi networks. Obviously Wi-Fi signals can go through walls, though they are attenuated in the process, and they are affected by other solid objects including people.

In a sense this is an old idea, as the original work on radar in the 1930s was inspired by engineers noticing that radio signals were affected by planes flying past. Professor Karl Woodbridge and his colleagues at University College, London, have shown how it is possible to "uncooperatively and covertly" detect people moving from the other side of a wall. (12) Woodbridge says a more refined version of the technology could detect breathing, so even staying still would not conceal you from it. MIT researchers are looking at a similar system that they call Wi-Vi; this is subtle enough to capture not just a moving person but to interpret their gestures.

Woodbridge's current system is the size of a laptop, but as usual, this is likely to follow the path of miniaturization dictated by Moore's Law. And since smartphones also have Wi-Fi antenna and the requisite processing power, it's easy to see how a phone-based drone, or perhaps several of them, might one day be able to scan a building for occupants remotely.

Again, it is a function that works well in combination with perching. A drone sitting unobtrusively on the roof could monitor comings and goings, keeping track of the number of people in a building at any time. It could make the difference between hitting an empty building and hitting the one occupied by insurgents next to it – or allow a strike to be called off if there are dozens of other people in the building besides the intended target.

Drones in the civilian world

A decade ago there was no real market for civilian drones. That has changed, and there are now a wide range of machines from radio-controlled children's toys to high-spec machines used by professionals in the movie business and elsewhere.

Progress has been extremely rapid. Some of the technology involved stems from advances in the smartphone industry, which we will explore in the next section. It is also driven by a growing market with a big consumer pull of its own.

The Micro Drone 2.0 that I encountered at the 2014 Farnborough Airshow has now been replaced with a version 3.0. (13) The camera has been updated to HD, trebling the number of pixels, and it now sends streaming video. More surprisingly, the camera is a gimbal mount, described as the smallest in the world. This is a feature that has only been available for the Raven since 2012. The gimbal means that, like the Raven, the Micro Drone can now swivel its camera to follow an object without having to change its course and can stay focused on a target while maneuvering in any direction.

As well as leveling out when tossed into the air, the Micro Drone can also fly upside down and perform aerobatics. A combination of sensors and automatic control algorithms keep it stable even in a crosswind. Its motors have been uprated, with a claimed top speed of 45 mph.

The Micro Drone is still marketed as a toy and priced at around $150. Its battery life of around eight minutes make it unsuitable for battlefield use, but the electronics could easily be grafted onto a bigger machine with a longer flight time and larger payload. Given that the electronics are the costliest part, a sub-$1,000 military drone looks like a real possibility.

The Micro Drone does not yet have a "follow me" capability, but there are other new drones – known as "selfie drones" – which can automatically follow the operator around and film them. They are intended for action sports like skating and snowboarding, giving a third-person view rather than the first-person view of body-mounted cameras like the GoPro. The Lily quadrotor costs $400 and follows you for twenty minutes. Again, the same technology could be ported to a larger machine to follow someone for military purposes.

Two: Phone Technology

It is a sobering thought that the first iPhone was launched as recently as 2007; the total number of smartphones has gone from about twenty million that year to around two billion in 2015. By some measures, the smartphone has been the fastest growing technology in history. It took landlines forty-five years to reach half of US households; smartphones achieved the same penetration in seven years, and during a recession. It is a phenomenally profitable business, as Apple's record-breaking $18 billion profit for the last quarter of 2014 showed.

The smartphone industry is driven by innovation. Apple and Samsung each reportedly spent some $14 billion on R&D in 2014 in their quest for technological supremacy. Progress has been relentless and looks set to continue even in the face of the laws of physics.

More than Moore's Law

People have been predicting the death of Moore's Law for decades. Various physical limitations looked set to slow down progress beyond the next few generations. Now we may really have encountered a show-stopper. Some current devices are only about a hundred atoms across, and are clearly approaching ultimate limits. On the current scale of operations, heat dispersal is becoming a major issue and will get worse with smaller components.

While Moore's Law may continue to find ways around these limitations, this may not be important. Smarter thinking may start to become more important than better hardware. Algorithms – the techniques used to solve problems automatically – are steadily improving. These developments rarely get any attention, even though they play a vital role in our technology.

For example, the unglamorous H264 data compression standard introduced in 2003 compresses video streams about twice as effectively as the previous H263 from 1994. The new HEVC (High Efficiency Video Coding) standard that replaced it in 2013 is twice as fast as H264. You can now transmit four times as much data as in 1994 over the same bandwidth thanks to better software. It is like quadrupling the speed of your car without upgrading the engine, just driving differently. This improvement is highly relevant in the world of drones, which beam back video and other data.

Similarly, few people outside the electronics industry have heard of the fast Fourier transform, but it has been a powerful tool since it was developed in the 1960s. The fast Fourier transform is a mathematical process used in converting signals from digital to analog and back. Its most important applications are in image compression – the technology that allows digital cameras, videos, and televisions to operate – and in the digital audio processing functions in digital phones, MP3 players, and radios. In 2013, a team at MIT unveiled a new "faster Fourier transform" that does the same job even more quickly, in some cases ten times as fast (14). Again, this will enable a new generation of software that runs significantly faster without upgrading hardware.

In other areas, from mathematical optimization to linear algebra, the progress has been faster, often well ahead of Moore's Law. Users are rarely aware of better software, and are far more likely to notice "software bloat" – the increasing inefficiency of software over time, sometimes to the point of canceling out the advances in processors. Twenty years ago, personal computer software had to be crammed on to just one megabyte of memory, as most computers only had one megabyte of RAM. The same word processing and spreadsheet tasks now take 256 megabytes minimum. Windows 95 required four megabytes of RAM to run, Windows 2000 increased this to thirty-two Mb and if you have Windows 8 you need a majestic thousand megabytes simply to turn on your computer.

Bloat happens when programmers waste extra power on prettier interfaces and features that are irrelevant for most users. This masks the steady progress of algorithmic improvement. Where it matters, the algorithms are outstripping Moore's Law in their rate of progress.

The effect can be seen clearly in niche areas of programming. When IBM's Deep Blue took on world chess champion Garry Kasparov in 1997, the IBM team needed every scrap of available computing power available to have any chance of winning. Deep Blue ended up with a purpose-built Frankenstein monster of a computer with thirty separate computing nodes and four hundred special-purpose chess processors.(15) Taking on a human chess player took one of the most powerful computers in the world at the time. This brute force computing approach allowed Deep Blue to look around six to eight moves ahead. In a closely-fought contest, the machine eventually defeated Kasparov by 3 1/2 games to 2 1/2.

Six years later, Kasparov was involved in another contest of man versus machine. This time he played against Deep Blue's successor, Deep Junior. The result was a drawn series at three games all. The biggest difference was that Deep Junior ran on a machine with about one per cent of the computing power of Deep Blue. Chess-playing algorithms had improved to the point of achieving virtually the same result with a hundred times less computing power.

Since then the chess-playing Mechanical Turks have continued to improve. Humans have not beaten machines at the highest levels since 2005. Another of Deep Blue's descendants, Deep Fritz, runs on Windows and other consumer platforms. By 2010, Deep Fritz had an Elo rating (the international rating for chess programs) somewhat higher than Deep Junior. It is also higher than the estimated rating for Kasparov. Since then Deep Fritz has been overtaken by other software. At the time of writing, am open-source computer program called Stockfish has the top rating. The massive hardware needed by Deep Blue is no longer necessary: Stockfish can achieve a higher rating than Kasparov with a high-end PC. (16)

Another landmark was reached in 2009 when a chess program running on a smartphone reached grandmaster level. Pocket Fritz only needs a fraction of the processing power of Deep Blue yet, thanks to improved algorithms, it has a superior rating. Unless you are one of a tiny number of individuals – literally less than one in a million – your phone much better than you are in the definitive game of intellectual prowess. Those few grandmasters are unlikely to enjoy their special status for long; the day when a smartphone program is ranked higher than any human chess player cannot be far away. Yet less than fifty years ago, Dr. Hubert Dreyfus, a philosophy professor at Berkeley, confidently predicted that no computer would ever rise to the level of beating a ten-year-old at chess.

Chess is a game of war, with pieces originally representing infantry, cavalry and chariots. Tactical combat may be more complex than chess, but the rate of advance in chess-playing software is a powerful lesson. Smart programming could turn drones with mobile-phone processors into tactical geniuses able to react quickly and out-think mere human opponents.

Chess is a good concrete example, but not the most dramatic case of algorithmic improvement. Professor Martin Grötschel of Konrad-Zuse Centre for Information Technology notes that one particular planning task used as a benchmark would have taken eighty years to solve in 1988 with the technology of the day. By 2003, processors were about a thousand times faster, but the algorithms used were now forty thousand times faster. The two have a multiplicative effect, so the process is now forty million times faster and the planning task can now be completed in less than a minute. (17)

Many of the challenges and limitations that face small drones are a matter of computing power and bandwidth. In Chapter 4, we saw that the Razor team still uses a commercial autopilot because the available smartphone processors and software could not react fast enough to fly the plane in real time. (Low-cost, open-source autopilots like Ardupilot at less than $100 are also getting increasingly capable.) Bandwidth is also a problem; it would be better to have a drone that did not send endless gigabytes of empty roads and streets and only transmitted information that would be useful, and doing this this would require more on-board processing.

Algorithmic improvement means that better performance is only a software upgrade away. And as the Razor team found, there is a big enough Android app marketplace so that any seller can find buyers and vice versa. Drones like Predator cannot join in because they are built around closed, proprietary architecture. But smartphones and smart drones will continue to get better, even if the hardware stays the same.

Thinking Cameras

One of the most striking areas of smartphone improvement is in photography, both still and video. Developers of phone cameras are faced with the same size, weight, and power challenges as the military for drone cameras. The difference is that phone camera developers have bigger budgets and tens of thousands of times as many users.

The first generation iPhone, back in the primeval mists of 2007, had a two-megapixel camera with no flash or autofocus. It could not shoot video. Seven years later, the iPhone 6 can record HD video at sixty frames a second or take twelve-megapixel stills – and can even take stills while recording video. Apple has run an award-winning advertising campaign called "Shot on iPhone 6" with images blown up to poster size to emphasize their high quality. This improvement has been achieved by a combination of hardware and software enhancements.

Smartphones are not generally equipped with an optical zoom because of the size, cost, and mechanical complexity this entails. Phones are generally stuck with digital zoom, and the loss of image quality that goes with it. Zooming by a factor of two means there are only a quarter as many pixels in the end result with a corresponding lack of fine detail. Even with millions of pixels, digital zoom is severely limited.

The size of the lens is a fundamental limitation. The laws of physics require that a small lens cannot resolve below a certain size. Camera makers counter this with computational photography, using processing to enhance the picture or video after it has been taken. Image processing technology can improve image quality, removing the blur and fuzziness and extracting sharp images. Similar techniques have been used by the military for decades; some of the earliest digital image processing was carried out by analysts working with the National Reconnaissance Office's spy satellites.

One of the leaders is Qualcomm, a maker of smartphone components including the Snapdragon range of image processors. (18) Apple's iPhones also have a dedicated image signal processor, part of the new Apple A-series chipset. Both have a handy selection of tricks that can only be carried out with the aid of multiple exposures and some serious computing power.

Qualcomm's design improves the effective resolution of the image via techniques collectively known as super resolution. One of the simplest forms is noise reduction – the camera compares several frames and averages them out, removing the random spatter of visual background noise to produce a clean image. Deblurring is a mathematical technique for converting a blurred image into a sharp one. It involves identifying bands of grey going from light to dark – the result of a boundary being blurred – and converting them back into sharp black/white boundaries. Other techniques look for features in the image to determine whether something spread over multiple pixels is genuine, or whether it is a blurred object that should be in a single pixel, and tidying it accordingly.

Super resolution currently gives up to x4 zoom with no loss of image quality. Other tricks include using what is known as "side information" – data about what the lighting is – to enhance images.

Qualcomm's computational photography also includes "Ubi-Focus," which produces a depth map of a scene and adjusts the images accordingly, so it can achieve the otherwise impossible feat of having every part of a scene in perfect focus. The same set of tools also provides a "touch to track" function, which automatically focuses on a specified object at a tap of the screen.

Both setups feature a high dynamic range or HDR mode. This takes multiple images with different exposures and combines them. If you are shooting a scene with a mixture of sun and shade, you normally have a choice of which of the two will be visible. If your exposure matches the sunny part, then the shade will disappear into uniform darkness; set the exposure for shade, and the sunlit section will be a whiteout. HDR combines the two pictures so you can see everything clearly; it is good enough that some camera experts recommend using it all the time. HDR is a neat trick, and one that the human eye cannot copy. It is likely to be ubiquitous in the snapshots of the future. It will make hiding in the shadows more difficult when the drones are looking for you.

Digital stabilization is common in digital cameras, so shots are no longer ruined by a shaky hand. The iPhone Image Processing System has automated image stabilization that takes four photos in quick succession, identifies which areas are sharpest in each of the four pictures, and combines the sharp sections into a single seamless image.

Qualcomm is taking stabilization to a new level by connecting the camera to the smartphone's gyroscopic movement sensors. The latest Snapdragon does not just detect shake after the event by the movement in photographs; instead, it anticipates the shake and digitally cancels out motion blur.

Similar tricks are used to stabilize pictures taken with long exposure, in a technique known as oversampling. This reduces the pixel count of the image but digitally reduces the "noise" created by long exposure. Smartphone cameras used to perform badly in low-light conditions; fierce competition has means they now produce crisp, high-quality pictures in even in dim light.

Developers have still more tricks up their sleeve. Apple has been rumored for some time to be working on a multi-lens system that combines pictures to produce images that are of quality comparable to a digital SLR camera. It would raise the bar for smartphone cameras, and provide drones with better imaging than ever.

The significance of this sort of development becomes more obvious when you compare it to the military sector. Video and Image Enhancement Workbench for Aerial Surveillance and Tracking (VIEW-FAST) is a set of software tools under development by Charles River Analytics of Cambridge, MA, for small drones under a USAF contract. VIEW-FAST comprises four modules with functions that may seem familiar: image correction to correct blur and improve contrast, super-resolution to reveal fine detail, target tracking to reduce the burden on the operator's attention, and image stabilization to compensate for camera shake. (19)

Small military drones do not seem to use image stabilization at present, so they are already a couple of generations behind smartphone cameras. In short, it looks like the Air Force's requirements for drones will deliver remarkably similar products to the smartphone industry. Except that given the smartphone's industry's bigger budget, they are likely to end up with a superior product than the Air Force.

Mosaicing is the process of taking a large number of pictures and stitching them together into one big picture. Many smartphones already allow users to shoot panoramas by sweeping the camera around, capturing the equivalent of several stills in one pass. There is clever stuff going on behind the scenes: gyroscope data is used to calculate what direction the camera is pointing for each frame, and the powerful processor orients and blends the images. This is the sort of task that MITRE needed a third-party app for when they carried out their aerial mapping demonstration. Now the capability is built into the camera.

The military and consumer requirements are converging in other areas. Qualcomm's computer vision team is working on 3-D object detection as a means of enhancing augmented reality apps. This is a "feature-based object detection solution," which identifies and locates objects based on particular features. For example, it might recognize a wheel and start looking for more wheels; when it has two or three is can start looking for the outline of a vehicle. Once it has tentatively sketched the outline, it compares it to a database of 3-D objects and identifies the vehicle, and, equally important, where it is and what its orientation is compared to the camera.

Amazon's unsuccessful Fire Phone featured an early version of object recognition called Firefly, thanks to a Qualcomm Snapdragon processor. Amazon's aim was to turn the world around you into a shop window: if you saw a jar of peanut butter, a painting, or a book, your Fire Phone could recognize it and order it from Amazon online at the click of a button. In practice it did not prove very effective, but it indicated the direction of travel of smartphone technology.

Object detection is essentially the same as a particular requirement that the Air Force has been working on for years. The biggest difference is that in their world it is called automated target detection and recognition. Rather than recognizing different types of consumer goods, they want to distinguish tanks, personnel carriers, and mobile artillery. A version of Firefly with 3-D object detection might do just that – as well as, for example, identifying high-value targets by their face or the sound of their voice. Such an app could be easily added to a small drone.

Qualcomm is also working on technology for SLAM, or Simultaneous Location and Mapping (20). SLAM is widely used in robotics to help machines find their way around unfamiliar spaces when there is no GPS or other external help. It uses sensor inputs to build up a map of the surroundings and navigate around them. Qualcomm's plan is to use SLAM to aid augmented reality apps; the phone has to know exactly where the camera is pointed to generate the correct display. Their approach picks out elements of the scene and builds a complete 3-D mesh model of it, updating at thirty frames a second even when the image is blurred, with "a state-of-the-art relocalization routine" so the phone can find itself if it gets lost.

This sort of SLAM would be useful for a drone would use in situations where satellite navigation was impossible. At the moment, even military drones like Raven do not have anything comparable; in the future it may be available as a free feature on smartphones.

Qualcomm is fully aware of the potential synergy between their chips and drones. They assisted Dr. Vijay Kumar at the University of Pennsylvania in using a Snapdragon chip to control a small drone: it analyzes images from its mobile phone camera to avoid obstacles and plot a path around them. (21)

Looking At You

Many smartphones' image signal processing software includes face detection. This is not the same as facial recognition, but it will identify the location of human faces in an image. This is something that people do naturally but that machine vision only accomplished in the 1980s after decades of research. It works by identifying what are known as adjacent rectangular areas with different levels of brightness – for example, the area with the eyes is almost invariably darker than the area of the cheeks. By scanning an entire image for patterns of rectangles of different scales, it is possible to spot faces with a fair degree of accuracy. This requires a lot of computing power – the sort of computing power that is now readily available.

These days phones do not just detect faces, they identifies the position of the eyes and mouth and the direction the face is pointing. It can then follow a face on the screen. This is handy for a photography app, as faces usually need to be in focus rather than the background. Of course, being able to detect and locate humans in a scene is also a very useful ability for military drones. It might be quite a challenging piece of technology to develop independently, but anyone with a smartphone can get their hands on it.

The field has moved on again. The latest iPhone -- like many others -- now comes with smile recognition and blink recognition. They now take multiple pictures and automatically select the one where people are smiling and do not have their eyes shut.

Beyond face detection is face recognition, a rapidly maturing technology. Interestingly, this is another area where the government used to have a lead but commercial concerns now dominate. The FBI's Next Generation identification system, deployed in 2014, has about an 85% success rate at determining whether two photos are of the same individual. Facebook, on the other hand, has software that is over 90% accurate – an impressive feat considering that Facebook pictures are not as precise as the standardized FBI mugshots. But then, Facebook is a bigger organization with more programming resources than the FBI, and has a larger database of photographs.

Meanwhile, facial recognition in portable devices is becoming more sophisticated and is starting to be used as a security feature. Facial recognition will never be enough for some purposes, particularly those involving security, so something better is required. The iPhone 6 introduced biometric recognition in the form of a fingerprint scanner, but the Fujitsu Arrows NX F-04G uses the phone's camera to scan the user's iris from normal reading distance, detecting the unique pattern in the colored part of the eye.

Previous devices had required the user to hold the scanner up to their eye and could take several seconds to work. The Arrows NX F-04G will supposedly provide foolproof, password-free user recognition. (22)

The US military has introduced biometric identification in areas troubled by insurgents. In Afghanistan, US and Afghan forces used handheld scanners to log iris patterns of hundreds of thousands of individuals. The database proved highly valuable.

"Simply put, biometric-enabled intelligence (BEI) efforts are producing a high return on operations designed to collect and exploit information about insurgents," according to a 2011 report, which lists several instances of suspicious individuals being picked up without ID and being identified as known insurgents by their iris scans.

A drone that can hover in front of someone and request them to look into the camera – backed up by other drones with lethal or nonlethal weapons – can act as an unmanned checkpoint, ensuring the identity of everyone passing is known. Anyone who resists is likely to be treated as hostile and the issue referred to a human controller.

Once, creating a drone with this sort of capability would have taken a major government contract; now it looks like a school science project. Currently, longer-range iris recognition cameras capable of "non-co-operative" identification already exist but are much bigger machines. With the steady advance of camera technology, smartphone-sized iris scanners are likely to improve.

Drones currently lack many of the features that we take for granted in phone cameras. If they do no more than catch up with where smartphones are now, it will be a huge advance. And by the time they have done that, smartphones will have moved on to the next level.

Nature's Technology

The advances promised by steadily improving electronics look impressive but less ambitious than those promised by the growing field of biomimetics, in which researchers set out to match nature's achievements.

Bug Eyes

Insects' eyes see differently to humans'. With a few exceptions, insects have compound eyes made up of hundreds of miniature eyes called ommatidia. This makes for poor long-distance vision, but an excellent wide-angle view. A fly has virtually 360-degree vision, so it is impossible to sneak up on a fly. Having many small windows on the world also makes it easy for insects to use what researchers term *optic flow*.

Optic flow is the way we see the world streaming past us. If you look out of a train window, you can see what direction the train is travelling from the movement of the scenery, even if the window is dirty or it is too dark to make out details. Flow not only tells you how fast you are going, it allows you to tell how far away things are: the further away an object is, the slower it moves. Distant mountains are almost stationary, houses next to the track rush past.

If you can see out the train windows in both directions, you can also see when the train is turning. If the track bends to the right, things move faster on the left side and slower on the right.

Optic flow tells an insect everything it needs to know as it flies through a cluttered three-dimensional environment. It is computationally highly efficient compared to the slow process of individually identifying objects. This is how a fly can easily spot a fast-moving object like a fly swatter moving towards it and take evasive action in a fraction of a second. Much of the computation for this takes place in the fly's visual system. In humans, the visual signals are routed through the nervous system and we are sluggish by comparison.

Centeye, based in Washington, DC, is a pioneering company in the field of applying optic flow to unmanned systems. Founded in 2000, the company had a series of firsts, culminating in 2009 with a system designed for DARPA to allow a small drone to hover in place. Their work is based on vision chips, which, like the insect eye, do their own rapid data processing. The eye fixes on a point on the ground and keeps it stationary, and that ensures that the drone is anchored in place. The entire system, including cameras, weighed under five grams and allowed the drone to hover using cameras alone, without gyros or other conventional stabilizers. By 2011 Centeye's researchers were able to do the same thing with a system weighing less than a gram. (23)

Centeye is now extending this for flight in the dark. Based on observations that many insects can fly around in what looks like pitch darkness to humans, they are designing a miniature guidance system for low light levels. For situations where there is literally no light, such as exploring underground, a single low-power LED provides all the light it needs while remaining virtually invisible to the human eye.

Fly Like a Moth

It is not immediately obvious how a quadrotor drone can imitate the flight of a hawk moth, but the InstantEye drone is an early example of a biomimetic drone. Flying in strong winds and cluttered environments is a challenge for small craft, but InstantEye copes with ease, maneuvering around with a precision that is little short of uncanny. On the outside it looks like other quadrotors, but it has a control system with flight rules derived from the behavior of moths, which gives it an unrivalled degree of stability and precision. (24)

The drone was developed by Physical Sciences Inc. of Andover, Massachusetts. The first step was to find out how moths manage to maintain their stability in all conditions. Moths were fitted with a tiny harness covered with reflective beads, and high-speed cameras recorded their flight, a motion tracking system similar to those used by filmmakers. The moths do collide with things, but they have a remarkable ability to recover. Usually they pitch up and push away from an obstacle so rapidly that the disturbance to their flight is too fast for the human eye to see.

"Typically they recover stability in about one wing beat," says Thomas Vanck of PSI.

The resulting algorithms were built into InstantEye, a drone that weighs barely a pound but can fly in winds of 35 mph and gusts of over 55 mph. PSI even successfully test-flew it in a hurricane. The Parrot AR Drone 2.0, a popular camera-carrying quadrotor controlled via Wi-Fi, is a similar size to the InstantEye but can only fly in winds of less than 10 mph.

"We have tested the system in very undesirable conditions – high winds, pouring rain, snow, sleet, blowing sand, and dust – and the system just takes it," says Vanck. (25)

InstantEye has a notably fast launch. When tossed into the air it climbs to four hundred feet in thirty seconds. Its performance indoors is equally impressive. InstantEye can easily negotiate confined spaces and slip through narrow doorways and windows.

Piloting is intuitive, with the autopilot doing all the tricky stuff. The operator pushes their viewpoint around in three-dimensional space, and the autopilot translates this into smooth motion. When not being controlled, InstantEye hovers as if anchored in space.

InstantEye has a camera that locks on to the surface below, allowing it to hover within inches of an object if needed, an impressive example of optic flow sensing in action. Like the hawk moth, if InstantEye has a collision, it recovers rapidly rather than crashing. The team has demonstrated its capability in a project that had InstantEye inspecting electrical towers. It provided close-up images of insulators, connectors, and other components sufficiently detailed to make human inspection unnecessary.

Unlike other quadrotors, InstantEye does not require expensive camera stabilization.

"Our autopilot is so fast and precise that it allows the whole camera to remain steady even when the weather is frightful," says Vanck.

Perhaps the most notable feature of InstantEye is its price: $6,200 for a setup that includes two air vehicles, a ground control station, and all the accessories. The advanced part is the software, which costs nothing to manufacture.

A smart, biomimetic design may be cheaper because it does not require add-on features like stabilization. Prices are likely to drop further if the model goes into mass production, as seems likely. Ultimately, this sort of intelligent technology could end up driving something like the $150 Crazyflie – which is, after all, even closer in size to the hawk moths that InstantEye is based on.

Brains for Drones

Perhaps the extreme form of biomimetics involves drones that think like living beings.

Flies, dragonflies, and bees can traverse obstacle-cluttered spaces and react with lightning speed to potential threats. Conventional computers have a digital architecture that works well for linear number-crunching but they are poor for tasks with a large number of inputs, and they cannot handle the sort of tasks that flying insects do all the time. In contrast, the neural network architecture of biological nervous systems is weak at calculation but excels in reacting to complex sensory input – and uses a tiny amount of power. The good performance of biological systems has spurred research into artificial brains that mimic nature's architecture, a field known as neuromorphic engineering.

The most common way to develop neural networks is with a digital computer with a set of software nodes or artificial neurons connected together. Rather than having digital ones and zeroes, it stores information in the form of the strength of the connections in the network.

For example, HRL Laboratories of Malibu is working on neuromorphic systems to improve small drones' ability to recognize objects, taking them to far higher levels than Qualcomm's Snapdragon. Their design is known as NEOVUS, for NEurOmorphic Understanding of Scenes. It takes the input from a five-megapixel camera at thirty frames per second and processes it in two stages. The first stage picks out objects of interest in the scene and passes them to the second stage, which classifies and identifies the objects.

NEOVUS can detect moving and stationary humans and vehicles in a scene, and even distinguish vehicles by type, telling cars from trucks. The technology potentially offers huge benefits to drone operators. Instead of having to watch the video feed the entire time, the human operator only needs to be involved when a vehicle appears on a road, for example, or people are seen exiting a building. It only needs a tiny fraction of the bandwidth of a system that, at present, sends back a constant stream of high-resolution video. (26)

The developers claim that NEOVUS uses less than a thousandth of the power of conventional computer vision algorithms, and the processing system is correspondingly compact. HRL has flown neuromorphic systems on a Raven, which gives an indication of the market they have their eyes on.

However, this is only a simulated neural network. The more radical approach is to build the real thing: a computer chip that is not digital but has the same sort of architecture as a living brain. This is exactly what Bio Inspired Technologies of Boise, ID, has done. Their work is based on a new type of electrical component called a memristor, a resistor with memory.

Bio Inspired's memristor is a tiny wire about a hundredth the thickness of a human hair, a wire that changes length when electricity is applied to it. The longer the wire, the lower the electrical resistance. When power is removed, the connection stays exactly as it is. The resistance is determined by not only the signal it is currently getting, but also the sum of all signals it has experienced in the past. The signals can complement one another, oppose one another, or combine in a complex fashion. In this respect, it closely resembles a familiar living structure.

"The fascinating aspect of the memristor, and one we are capitalizing on, is the uncanny similarity of this behavior to the biological synapse," says Bio Inspired CEO, Terry Gafron. (27)

The company has a range of memristor products coming out under the name Neuro-Core. The most advanced is the beta version of a memristor processor chip (28). To start with, these will be used as more efficient hardware alternatives to the sort of digital simulations of neural networks carried out by NEOVUS.

Neuromorphic hardware may only have a handful of elements compared to the millions of transistors on a normal chip, but as

Gafron points out, a fly's complex flight requires only a few hundred neurons.

"Can a bee-sized brain control a drone? The answer is an emphatic yes," says Gafron, and compares insect brains with the digital flight computers currently in use.

"The F-35 uses a handful of sensors and three computers the size of shoe boxes to control flight and still can't outperform the fly for maneuverability, power consumption, or efficiency. If memristors emulate the biological synapse, they can be used as flight control systems that approach insect performance levels. Our Neuro-Core line of products is a move in this direction."

Gafron has been "training" neuromorphic hardware to recognize objects like trees and planes, sometimes by literally showing models of them to a camera linked to the device in the lab. The neural network soon learns to pick out the features that distinguish a plane from a cloud, and can go on to identify types of aircraft from their shapes.

It is not just the speed and efficiency that makes neuromorphic computers so effective. Gafron says that the Holy Grail for this work is to create computers that "learn." A drone could start out with basic programming that allows it to fly, but the neuromorphic brain would steadily get better with experience like a human pilot.

"You could have battle-hardened drones teaching others how to operate," says Gafron.

Once one drone learns how to perform a maneuver, the knowledge could be shared over a network. The more neuromorphic drones, the more learning and the more new tricks there are to share. The entire drone fleet would have the benefit of millions of hours of flying experience in all weather and conditions, in every sort of terrain. Neuromorphic systems can learn from their mistakes and correct their own weaknesses. Every drone would be piloted by an ace.

Gafron expects the neuromorphic brains to progress roughly in line with Moore's Law, with successive chips getting progressively smarter and more powerful. Nobody is talking about human-level intelligence yet, but the history of previous computers points to a rapid and continuing upward evolution.

As this chapter has shown, future small drones are will be progressively more capable than the current versions. While there is no way of knowing which of the technologies here will be matured and which will fall by the wayside, the progress in each area suggests that new devices, such as LADAR, laser designators, and radar that can see through walls area likely. There will certainly be camera systems with ever greater capability to distinguish and recognize objects. New drones may be as agile and capable as birds or insects, and possess a level of autonomous intelligence which is hard to imagine.

The new drones will also be cheap, which presents challenges in how to control the large number of drones. The answer to that lies in the science of the swarms.

REFERENCES

1) DJI Matrice 100 (n.d.). (http://bit.ly/1QWRsGc)

2) MAV6 LLC Raven upgrades. Mav6. (n.d.).
(http://bit.ly/1O63VBo)

3) Scientific Applications & Research Associates electronic warfare

Small UAV accurate geolocation and discrimination. (n.d.).
(http://bit.ly/1LRfbzE)

4) Wireless Aerial Surveillance Platform (n.d.).

5) AIRTRAC laser designator

"We developed a proprietary three-dimensional resonator" – James
Murray of AIRTRAC interview, with author July 2014

6) Elbit Rattler (n.d.). (http://bit.ly/1lZYbTa)

7) Velodyne LIDAR (n.d.). (http://bit.ly/1Qie8iw)

8) Steerable Electro-Evanescent Optical Refractor

Compact 2D EO 1.55 µm Laser Scanner. (n.d.).
(http://bit.ly/1LRfcUn)

9) SEEOR

The Economist. (2012, November). Laser sight [Web log post].
(http://econ.st/1PB6bqf)

Hambling, David. "Developers eye mini-lasers for small UAVs."
Aviation Week.

10) Kenai micro radar

Marshall, P. (2014). Miniature radar may put UAVs in the air [Web
log post]. (http://bit.ly/1jBsvS6)

11) Camero "Airborne structure penetrating life detection system."

Camero launches unmanned VTOL structure-penetrating life
detection system. (2013, November 19). (http://bit.ly/1QWRWw7)

12) Karl Woodbridge Wi-Fi Radar

Hambling, D. (2012, July). Seeing Through Walls With a Wireless
Router. *Popular science*. (http://bit.ly/1jBsAVY)

13) Micro Drone 3.0

Micro Drone 3.0: Flight in the palm of your hand. (n.d.). (http://bit.ly/1YLDGrt)

14) Faster Fourier Transform

Hardesty, L. (2012). The Faster-than-Fast Fourier transform. *MIT News*. (http://bit.ly/1QieCVY)

15) Deep Blue and improvements in computer chess-playing

Richards, M. & Shaw, G. (n.d.). Chips architectures and algorithms. (http://b.gatech.edu/1jBsJZv)

16) "Since then it has been overtaken by other software"

Stockfish CCRL 40/40 computer chess rating. (n.d.). (http://bit.ly/1TnNZy9)

17) Professor Martin Grötschel of Konrad-Zuse centre for information technology notes…

Nisan, N. (2010, December 23). Progress in algorithms beats Moore's law. [Web log post]. *Turing's Invisible Hand.* (http://bit.ly/1ot5UsG)

18) Snapdragon Computational Photography

Qualcomm technologies announces next generation qualcomm snapdragon 805 "Ultra HD" processor. ((2013, November 20). (http://bit.ly/1OJGRuP)

19) VIEW-FAST

Video and image enhancement workbench for aerial surveillance and tracking (VIEW-FAST.). (n.d.). (http://bit.ly/1Q3w6a5)

20) Qualcomm SLAM

Qualcomm research office projects page. (n.d.). (http://bit.ly/1MZ2ruI)

21) Qualcomm Snapdragon as drone guidance

Matthews, L. (n.d.). Quadcopter becomes autonomous by using your smartphone as a brain. (http://bit.ly/1LRfdYj)

22) Fujitsu Arrows with Iris identification

Fujitsu releases ARROWS NX F-04G equipped with functions such as quick and easy-to-use iris authentication. (2015). (http://bit.ly/1HIXR5d)

23) Centeye optical flow. Centeye. (n.d.). (http://bit.ly/1XCVuYo)

24) InstantEye biomimetic moth drone

Hambling, D. (n.d.). Moth drone stays rock steady in gale-force winds. *New Scientist.* (http://bit.ly/1lcqYnh)

25) "Our autopilot is so fast and precise ..." Thomas Vanck interview with author, January 2014

26) HRL Neuromorphic Visual Understanding of Scenes NEOVUS

Khosla, D. (n.d.). A neuromorphic system for object recognition. (http://1.usa.gov/1Rqfp86)

27) "The fascinating aspect of the memristor," –Terry Gafron interview with author, July 2014

28) Bio-Inspired neural memristors (http://bit.ly/1lcr7Y1)

CHAPTER SEVEN

TEAMS, FORMATIONS, AND FLOCKS: THE POWER OF THE SWARM

*And what's more wonderful, when big loads foil
One ant or two to carry, quickly then
A swarm flock round to help their fellow-men.*

- John Clare "The Ants"

The future will bring new generations of increasingly capable small drones at ever-lower costs. Large numbers of them will be more formidable opponents than the handful of armed Predator/Reaper drones currently in service. They will be more formidable still because they will not be operating independently but working together in a swarm that is greater than the sum of its parts.

The term *swarm* is used in this book to describe drones working cooperatively together. It is commonly used in robotics, but not everyone likes the word, which has some negative connotations. The Pentagon uses *swarm* to describe opponents, such as the swarming boat attacks practiced by the Iranian Republican Guard. Although *swarm* suggests drones flying like bees in a dense formation, a swarm can be dispersed over a wide area. What distinguishes a swarm from a collection of individuals is that the elements work together as a team. This process requires minimal brainpower on the part of the swarm elements, but it allows control to be applied at a swarm level rather than an individual level. One human operator can handle many drones as a swarm, or the swarm can act as a single autonomous unit without human control.

The Expendables

With rudimentary cooperation, a gang will always overwhelm a single opponent. Hollywood heroics notwithstanding, numbers win.

"Quantity has a quality all its own," Stalin is said to have declared. His Red Army crushed the Wehrmacht veterans by weight of numbers. The Germans aimed to field the best possible equipment with a high standard of engineering; the Soviet approach was to produce simple, well-designed but crudely finished hardware in massive quantities. Their tanks, aircraft, and artillery were basic but effective and were churned out in sufficient numbers to ensure victory.

Consider two batteries of artillery firing at each other. Obviously enough, the one with twice as many guns fires twice as many shells. Less obviously, it only has half the number of targets to destroy before the enemy battery is silenced, and this has a multiplicative effect with the first factor. The outnumbered battery must be four times as efficient to have a chance of winning. In military analysis circles, this is described as Lanchester's Square Law: military power is related to the square of the number of units.

This approach is dependent on a willingness to sacrifice numbers of your own troops. The Russian military took huge casualties during WWII, with something over eight million dead, but the German invaders were relentlessly ground down and ultimately defeated. The Russians could not match the Germans battalion for battalion, but by having more troops and being willing to accept twice as many casualties as they inflicted, they eventually won.

As we saw in Chapter 5, large numbers and low quality is not an approach that sits comfortably with modern Western culture. There is no such thing as an acceptable level of casualties. "I will never leave a fallen comrade," is part of the US Soldier's Creed. The Pentagon aims to provide the best possible equipment, giving every soldier and pilot the best possible chance of survival. Soldiers are not sacrificed lightly for the sake of a quick victory.

However, this only applies when it comes to human lives. Robots are different. The ability to absorb casualties becomes a key asset.

"The principle advantage of a swarm is robustness," says Stephen Crampton, CEO of British company Swarm Systems. They have developed flocks of small quadrotor helicopters called Owls for military missions like urban reconnaissance. "If eight vehicles go out and two are lost, then the other six can reform to carry out the task."

Unlike humans, losses mean nothing to the swarm. There have been many studies of how soldiers behave on the battlefield. While there are heroic exceptions, the rule of thumb is that the morale of military units cracks on the battlefield when casualties reach 25-35%. When between a quarter and a third of the troops are killed or incapacitated, the attack tends to falter, or the defenders start retreating or surrendering. The situations where this does not happen, where elite units or fanatics fight to the last man, tend to be the stuff of history and legend.

With drones, casualties are irrelevant. A drone swarm which has lost 50% or even 90% of its members will not be disheartened. This unblinking resistance makes them terrifying opponents. Drone swarms might resemble advancing Zulu hordes or Chinese "human wave" infantry attacks that storm forwards in spite of any casualties. The courage of the famous three hundred Spartans who fought against the Persian host at Thermopylae was remarkable enough to be remembered over two thousand years later. But by drone standards, it is routine. For humans it is disconcerting to face an enemy with no fear of death.

As Crampton notes, the ability to take casualties makes drone swarms robust compared to single large drones. One broken component can end a Reaper's mission, and one missile can knock it out of the sky. But a swarm remains a swarm and can continue on its way even after heavy losses.

This makes swarms especially dangerous in the context of air defenses, which have become highly reliant on single large missiles to knock out attackers.

As we have seen, the Naval Postgraduate School in Monterey, California, is a center for work on swarming drones. A 2012 study by Loc Pham called "UAV swarm attack: protection system alternatives for Destroyers" gives a sobering assessment of what would happen if a US Navy destroyer were attacked by a number of cheap, unsophisticated drones. These would be effectively "flying IEDs" making suicide runs.

The destroyer is protected by a system called Aegis, named after the legendary shield of Athena. Aegis is a sophisticated arrangement that coordinates various radars, guns, and missile launchers into a tightly integrated defense grid. It can detect, identify, track, and intercept incoming aircraft, missiles, and other threats in seconds, with minimal oversight from its human operators.

Aegis's defenses start with long-range Standard missiles, and end with the Phalanx Close-In Weapon System, a Dalek-like turret with a multi-barreled 20mm canon spitting out seventy-five rounds a second. Aegis is considered one of the most effective defensive systems in the world.

The Monterey researchers also assumed that Aegis would be supplemented by machine-guns manned by sailors on the destroyer's decks as a last-ditch defense. These were added because the Phalanx cannot engage anything closer than about two hundred yards away, and it was clear that some of the drones were going to get through.

The problem was that the suicide drones were so small they were often not detected until they were dangerously close. When they were spotted, the defenses only had a limited time to shoot down several difficult targets. After running a computer simulation five hundred times, the researchers concluded that with eight drones approaching simultaneously, four could be expected to get through and hit the destroyer.

Four small warheads would not sink a destroyer, but they could certainly damage it and cause loss of life. Those sailors out on the deck manning the machine guns might become casualties. Worse, if the attacking drones targeted vulnerable points like radar or missile launchers, they might leave the destroyer defenseless so that it could be sunk by larger weapons.

The study went on to look at various ways of improving the defenses, by having better radar or greater accuracy or longer range weapons. These improvements reduced the number of hits, but only up to a point. Even with all the improvements in place, the average number of hits was reduced from four to 1.3.

The researchers also found that the number of incoming drones made a bigger difference to the outcome. Even the baseline destroyer with no improvements could beat off an attack by five drones; ten would tend to overwhelm even the best defenses and score at least one hit.

Increasing the numbers of attackers increases the number of hits, and real-life opponents would not be limited to ten drones. The researchers did not look at what twenty or fifty drones might do, but the pattern is clear. Existing defenses can only knock down so many incoming, and the rest will get through. Hollywood is fond of the small band of heroes who defeat a vast army of opponents, which is why movies are still being made about the Spartans. In the real world, the large numbers are likely to win. Quantity really does have a quality of its own.

Spreading Out

The other benefit of a swarm over an individual is that they can be in more than one place. A 2014 study by RAND analyzed the effectiveness of different types of unmanned aircraft on a "hunter-killer" mission to find and destroy a moving vehicle within a large area. Their baseline was the RQ-9 Reaper, which they compared with other options.

The Reaper has powerful sensors, so it can fly at high altitude and sweep a wide area. Smaller drones have to fly lower to get as good a view with their cameras, so each drone covers a smaller area.

However, the overall advantage was with the smaller aircraft. The RAND study concluded that "two or three smaller RPAs [Remotely Piloted Aircraft – drones] with less-capable sensor packages were often able to equal or exceed the performance of the larger RPAs employed singly." Interestingly, as with the Monterey study, the RAND team did not include an option for a larger number of very small drones, but did at least note that the idea merited further research.

Stephen Crampton of Swarm Systems points out that the advantage goes up directly with the number of drones.

"Swarms reduce mission time," says Crampton, describing work showing that several drones will complete a job quicker than a single aircraft.

Covering a larger area, or covering a number of specific points scattered over a very wide area, gives the swarm an advantage. The advantage is even greater when they are able to use the same sort of swarming tactics as ants and termites, as we shall see.

Natural Born Swarmers

In technical terms, a swarm may be defined as a number of relatively unintelligent agents that all follow the same set of rules to achieve coordination. We see this in nature without realizing it. A flock of birds can maneuver in tight formation, navigate together, and even land on the same tree without any risk of collision. Fish and insects show similar swarming behaviors while traveling, foraging, or nest building. Their abilities sometimes seem to verge on the miraculous.

British ornithologist Edward Selous was of the first to make a serious study of flocking birds in the 1930s. He was fascinated by bird behavior and described in painstaking detail coordinated movement within flocks, including rooks, gulls, lapwings, dunlins, swans, and starlings. All of them seemed to work together as though they were parts of the same huge organism. Rooks rose from a field "like a leaping up of black flame, so instantaneous and unanimous was it. It is transfused thought, thought transference – collective thinking practically. What else can it be?"

The most dramatic example of bird flocking behavior is seen in starlings, which gather together in gigantic groups known as murmurations. Thousands of birds fly together, creating the impression of a cloud of smoke that constantly changes shape, as they wheel, turn, and swoop together. In spite of the sudden and rapid changes of direction, there are no air-to-air collisions. The whole thing is as perfectly choreographed as an air display, but utterly spontaneous and without a script.

Selous found the whole performance baffling. He came to the conclusion that it could only be explained in terms of telepathy within the flock. His 1931 book, wonderfully titled *Thought-Transference (or What?) in Birds* described the behavior and why it could not be achieved through normal senses. He started at the idea that there might be leaders within the flock that signaled actions to the others, but soon concluded that this could not be the case and that some sort of group-mind must be at work.

"Their little minds must act together. Though I cannot understand it, yet it seems to me that they must think collectively, all at the same time, or at least in streaks or patches—a square yard or so of idea, a flash out of so many brains."

Termites also seemed to have powers far beyond what could be expected of mindless, short-sighted insects. How do they manage to build something as complex as a termite mound – surely there is some leader directing their activity? Again, a hidden coordinating agent was the most likely explanation. South African biologist Eugène Marais published *Die Siel van die Mier* (*The Soul of the White Ant*) in 1937 in which he identified the queen termite as the motivating force:

"The queen is the psychological center of the community; she is the brain of the organism which we call a termitary. From this shapeless, immobile object, imprisoned in her narrow vault, there emanates a power which directs all the activities of her subjects."

Both biologists were mistaken, and the answer turned out to be far simpler than either imagined. There is no mental telepathy or unseen power involved. Just by responding to the motions of its immediate neighbors, each bird ensures that the entire flock of thousands is more tightly coordinated than an aerobatic display team. Similarly, the termite mound is built through each termite mindlessly following a fixed set of rules depending on what their neighbors are doing. Contrary to all appearances, there is no central organizing intelligence.

A key feature of swarms is that none of the members is in charge, and there is no external "brain" controlling them. Control is decentralized or distributed, and there is no leader whose loss will cause the swarm to lose cohesion. In this sense it has no vital organs. Although the entire swarm might be considered a single organism, unlike other organisms, it cannot be killed by destroying specific parts. Half a swarm is still a swarm and still capable of all the same actions.

Software simulations of flocks were pioneered by American artificial intelligence legend Craig Reynolds, who created computer simulations using beings called Boids – a word somewhere between Bird and Droid. In 1986 Reynolds showed that movement of groups of creatures in nature, from schools of fish to flocks of birds or swarms of insects, can be modeled by having each individual follow three rules:

1) Separate: Keep a certain minimum distance from your nearest neighbors.
2) Align: Steer towards the average heading of your neighbors at the same speed.
3) Cohere: Attempt to move toward the average position of your neighbors, keeping the flock together.

When combined with behaviors such as heading towards a goal or avoiding obstacles, Reynolds' computer-generated flocks of Boids maneuvered together just like their biological counterparts, without any central control. Swarming robots can move together with a fluid agility that human pilots might envy. The software provides convincing flocks of CGI birds and bats for Hollywood blockbusters like *Batman Begins* without having to specify a flight path for each member of the flock. In 1998 Reynolds was given a special Academy Award in recognition of the development of animation.

Hollywood now relies on swarming software to fill out scenes that might once have required thousands of extras. In Peter Jackson's epic adaptations of *The Lord of The Rings* and *The Hobbit*, vast armies of orcs, elves, and others clash in Middle Earth. Zoom in and you will see that each animated figure moves separately, maintaining distance from its neighbors while heading in the same general direction and staying together. The same three rules, with suitable extension, create a realistic impression of a living mass of beings. The alternative approach – shooting one group of extras and digitally copying them hundreds of times – looks lifeless and artificial by comparison.

The lack of central coordination makes swarm behavior attractive to drone developers. Rather than needing one operator per vehicle, the operator can set the goal for the group as a whole. Atair Aerospace pioneered this with their Onyx Precision Airdrop system developed for the US Army in 2008.

Onyx is an unmanned parafoil; once dropped from an aircraft at altitudes of over 20,000 feet, it steers itself and glides down to a precision landing at the target site on the ground with the aid of GPS. The Ultra-Light version can carry up to seven hundred pounds of cargo. The airdrop gets challenging when you have a large number of payloads all heading for the same drop zone. One Hercules transport can carry over forty thousand pounds of supplies for air drops, which would mean several dozen Onyx. And if some of the payloads dropped later are heavier and will reach the ground faster, there is a real risk that the parafoils are going to run into each other.

Atair avoided these problems with flocking/swarming algorithms that ensure that there are no collisions and the whole group is beautifully coordinated as if by telepathy. They may not be as spectacular as a murmuration of starlings, but the Onyx parafoils show that the same principle that works for birds works for unmanned vehicles. Without any human intervention, the whole flock finds it way home to roost.

Since then, swarming software for unmanned vehicles has become increasingly sophisticated. Corvus, developed by Axon AI in Harrisonburg, enables a single operator to control a swarm of drones as well as CCTV cameras and other devices from one dashboard. The drones are autonomous, and the software rearranges them according to need, like an advanced version of DARPA's HURT RSTI mentioned in the previous chapter. Corvus suggest that their software might be useful for search and rescue, or to patrol forests looking out for fires. They also mention security and rapid, stealthy reconnaissance on the battlefield. There does not seem to be a military market for this yet, but that may change soon.

The US Navy's Low-Cost UAV Swarming Technology (LOCUST) program is perhaps the most serious military swarm program to date. In summer 2016, the aim is to have thirty drones flying together without having to be individually controlled, maintaining separation safely like a flock of birds. These drones are Coyotes, similar in size to the Raven, but the sensors and software could work with any drones.

"Precise formation control of large numbers of UAVs while conducting various maneuvers is a major demonstration objective," says Dr. Lee Mastroianni, LOCUST's project manager.

And, rather than having each drone piloted individually as with every current drone from the Reaper to the Raven, the operator manages the swarm as a single unit.

"The Swarm Operator Interface is one of the significant breakthroughs in this effort," says Mastroianni. "Generally the entire swarm will be managed by a single operator."

However, he notes that the swarm does not need to be a single entity: it can break into multiple swarms, each controlled by an operator, or drones can break off to perform missions (scouting or attacking) autonomously. Swarm control will make drone piloting a very different experience.

The thirty-drone demonstration is just the start. Mastroianni is committed to low-cost drones, aircraft so cheap that they can be treated as expendable. The aim is to have as many of them as possible, and much larger swarms are likely to follow.

Swarm intelligence

Similar sets of rules to those for formation flying can produce efficient foraging and nest building. The secret of the termite's engineering skill was uncovered by French biologist Pierre-Paul Grasse in the 1950s. He found that if a termite carrying a pellet of chewed soil comes across a mound of pellets dropped by other termites, she will add hers to the pile. The grains, stuck together by termite saliva, gradually grow into pillars. When they reach a certain height, a different behavior kicks in and the termites start building sideways to join the pillars together. The works have no interaction with each other at all, reacting only to the growing piles of earth; the triggering factor is that they recognize the smell of other termite's saliva.

A similar process helps them repair breaches and maintain the complex "air conditioning" system that keeps cool, fresh air circulating in the nest, all the way down to its heart several feet below ground level. Their response to feeling outside air is to go and drop a grain of dirt at the entry point; thousands of workers do this, and the hole is quickly plugged. When there is no more hole to plug, the termites continue carrying their chewed earth around. If there is not enough air circulation, they start removing internal partitions to increase the flow; if it is too cold, they build more partitions. The termite mound is in a constant state of rebuilding.

The termite defense force consists of soldier termites with outsize jaws. If the mound is attacked – usually by predatory ants – these soldiers rush out and hurl themselves at the enemy to delay the attackers while workers seal up the entrances. The soldier termites are left on the outside and doomed. It may seem a callous strategy, with heroic individuals ruthlessly sacrificed, but it is an effective one and the colony survives. Stalin would no doubt have approved.

Termite workers also follow simple swarming rules. Individual termite activity looks purposeless and unguided: a termite may pick up a pellet of earth, move it, put it down, and then pick it up and move it again several times. They wander around seemingly at random. What you cannot see is that every termite leaves an invisible pheromone scent-trail that others can follow, and the sum of these trails directs their movement.

This turns out to be a brilliant solution to the problem of organization. If one termite locates a food source, the entire nest does not drop what it is doing and go out foraging, but the trail she has left means that it will be visited by further termites, and their trails will attract more foragers. When the food source is exhausted, the foragers start wandering off randomly and the trail gradually fades. In this way, the number of foragers is automatically adjusted to the size of the food source, with no external coordination needed.

(This, incidentally, leads to the easiest and most effective way of keeping ants out of your house. If you find a trail of foragers leading to an open jar of ant delicacy, just sweep them up, toss them back outside and wipe up the trail with a damp cloth. Without the trail of pheromones, no more ants come looking for food.)

While more termites will follow stronger trails, the randomness element and the number of termites ensure that they will not miss anything interesting. If a hungry crowd of humans sees an obvious food source, such as a burger stand, the entire group may end up standing in line. Some may wait a long time, even though there might be another burger stand just around the corner and two more on the next street. Termites are much better adapted to searching out and exploiting every available food source within range. They never get bored or try to be clever, but they find everything and home in on it with as many foragers as needed.

The overall effect of this apparently chaotic and unintelligent individual behavior is a highly effective, robust, and flexible organization. Termites have been around for over a hundred and forty million years, and their simple but effective strategies have been honed to a fine edge by evolution.

Similarly, when it comes to searching an area, a swarm of drones can be effective by following the same approach as foraging insects: randomly wandering about and then gradually clustering around the areas that look most interesting as they are identified.

Programmers have even developed sophisticated search techniques based on this principle. Particle swarm optimization has many software "agents" going through a search space. They might, for example, be looking through a frame of video to see if it contains a building. When one agent finds something that might be part of a building, for example a straight edge or corner, it summons others to the general area. The agents move away from the less-promising areas and concentrate where there are most building-like features. This approach is far more efficient than the intuitively obvious approach of scanning systematically through the frame line-by-line as a human analyst might tend to do.

Hunting in Packs

A swarm attack is not just a mindless surge in the general direction of the target. Nature provides many examples where swarms increase their hunting effectiveness by working together.

Wolves are unusual among carnivores in that, in some areas, they prey largely on animals larger than themselves. Not only are moose and bison several times bigger than wolves, they are also faster. But a pack of wolves can bring down a large prey animal by working in a pack and using a set of heuristics – simple hunting tactics from a combination of instinct and experience. Although the wolves may appear to be carrying out a clever ambush, research suggests they are working with just two simple rules resembling those controlling the Boids.

Raymond Coppinger and colleagues at Hampshire College in Amherst, Massachusetts, created a computer simulation of a pack of wolves stalking a large prey animal. During the approach, each wolf moved towards the prey until it reached a certain distance; it then moved away from any other wolves that were the same distance. The net effect was that the wolf pack spread out and enveloped the prey. If the prey tries to circle around, the pack keeps homing on it, and in simulations the prey often ended up running towards one of the pursuers and was "ambushed" by it. Even though the prey may be faster than the wolves, it keeps turning to get away from the nearest wolf. This zig-zagging slows it down so that other member of the pack travelling in a straight line can catch it.

These simple tactics do not need elaborate communication between pack members; all the wolves need to know is where the next nearest wolf is. Their barks as they chase the prey can simply convey "I'm here."

"The form of cooperation needed for wolf-pack hunting can take place even among strangers," Coppinger notes. "Seemingly complex collective behaviors can emerge from simple rules, among agents that need not have significant cognitive skills or social organization."

Harris hawks are medium-sized hawks native to the Americas, found from the southwestern US to Chile and Argentina, which use a variety of approaches to attack prey. They are among the few birds of prey that work cooperatively, often in family groups of four to six birds. The most common tactic is a simultaneous attack with multiple Harris hawks diving in from different directions; a rabbit or other prey may dodge the first hawk or two before getting picked up by the third or fourth.

When prey goes to ground, the hawks switch tactics. The birds perch around the cover where their target has hidden, surrounding the prey, and then take turns attempting to penetrate the cover. As soon as the prey is flushed out, the surrounding birds swoop in and take it.

Finally, Harris hawks also carry out "relay attacks" in which multiple birds swoop down one after the other, each chasing the prey for a short distance. As it escapes one hawk, the next one in the flock takes over. Researchers have recorded up to twenty swoops in one chase over half a mile before the exhausted prey was finally taken.

Each of these approaches involves all the birds following the same set of rules. As with the wolves, they do not need to confer, and no leadership is needed, but they produce highly effective tactics.

Both the wolf pack and Harris hawk approaches are applicable to small drone swarm tactics. Without any need for a human to give them specific tactical instructions, a group of drones can effectively work together and become more effective through coordination.

The US Navy has demonstrated the swarming approach with robot boats known as Unmanned Surface Vessels or USVs. In 2014, a fleet of thirteen boats equipped with CARACaS (Control Architecture for Robotic Agent Command and Sensing) were set to escort a Navy ship. When they detected the approach of a potentially threatening enemy ship, the boats responded as a swarm. Each plotted its own route and coordinated with the others in collectively heading for the threat. The Navy has said that CARACaS boats could "detect, deter or destroy" threats, but noted that a human operator would need to approve any decision to open fire.

The USVs would be valuable for deterring suicide attacks from small vessels, like the one that killed seventeen sailors on the USS Cole in October 2000.

"If the USS Cole had been supported by autonomous USVs, they could have stopped that attack long before it got close to our brave men and women on board," Rear-Admiral Klunfer commented on the CARACaS demonstration.

Fly Together, Fly Further

While swarming drones may spread out over a wide area, they might also get some advantages by flying together in formation like geese and other migrating birds. Large birds are often seen flying in skeins, V-shaped formations, with the birds spaced at regular intervals. While early naturalists suggested that each flock had a leader who was guiding the rest, closer observation showed that the formation changes over time, with different birds taking the point position. It is all a matter of aerodynamics.

The tip of a wing, whether it is a goose or an Airbus 380, generates a whirlpool of air known as a tip vortex. This produces a downwash beneath the wings and an upwash just outside the wing. The vortex is actually a miniature tornado that may contain airspeeds of 100mph and may be about the same size of the span of the wing that produces it. The vortex from an airliner can be dangerous, as it is strong enough to flip a light aircraft right over. Close to the ground, the vortex from an aircraft taking off may persist for more than a minute.

Birds, being masters of riding the air, know how to take advances of vortices. By flying just to the side and behind, a following goose gets the benefit of the updraft provided by its companion. This gives it free lift, equivalent to flying downhill.

This is distinct from slipstreaming or drafting by cyclists to reduce air resistance, but it produces the same type of advantage. The overall benefit in range increases with the square root of the number of flyers in the formation. Naturalists' estimate that skeins of geese gain a 70% range advantage by flying in formation rather than individually. A detailed aerodynamic study by the US Air Force Air Vehicles Directorate found that formations of nine aircraft could achieve an 80% increase in range over the distance they could fly alone.

If human runners could gain a similar advantage, no long-distance event would ever be won by a single runner without a team. If you could gain similar benefits for the gas mileage in your car, the freeways would be a steady stream of cars in V-formation.

Manned aircraft avoid flying close together for safety reasons, and there are few situations in which aircraft travel in convoy. But swarms of drones might fly together in large numbers without such concerns. In addition, drones are not limited to the simple V-formation adopted by geese. Large numbers of drones could adopt more elaborate formations involving multiple V-shapes stacked to give greater overall benefits.

Formation flying might become the standard for drones crossing long distances, as well as for "eternal aircraft" that remain on station for weeks or months.

Radio Formation

There are other advantages for cooperative drones that are not available to birds. The work on radar for small drones mentioned in Chapter 5 raised the possibility of several drones working together to act as a single large radar dish, an arrangement known as a "sparse antenna array." The emissions from several drones are combined together into a single signal in a process called beamforming.

Christopher Kitts of Santa Clara University has been testing this approach as a way of boosting radio communication for drones. He has tested the theory in field trials with a squad of small ground robots. To form the sparse array antenna, the robots arrange themselves at roughly equal intervals, generally in a line or a circle. One robot acts as master and has software that takes information about the signal at each of the antennas – specifically the amplitude, phase, and delay – and combines this information to get an enhanced signal. There is little extra hardware needed as the system exists almost entirely in software, so any of the drones can be the master.

"Combining the multiple elements simply gives you something unobtainable from a single robot," says Kitts. "We might use a group of six 'bots to receive communications from a very distant entity. We could store this transmission, then rotate the formation, and relay the communication to a separate entity."

Alternatively, three of the 'bots can form an antenna in one direction and three in another direction, forming a relay bridge between two remote sites.

The technique is also useful for dealing with jamming, as Kitts' colleague Garret Okamoto of Adaptive Communications Research Inc. noted. In a process known as nulling, the jammer signal is detected and located from its influence at different points, then completely cancelled out.

"I showed that we could completely ignore the effect of a jammer, positioned between the transmitter and receiver, that was twenty-three decibels above the desired transmitted signal," says Okamoto.

Twenty-three decibels means the jammer is transmitting noise two hundred times as powerful as the source signal. It is like being able to hear a distant whisper while someone is bellowing in your ear.

"We could have nulled a stronger jammer," says Okamoto. "But we couldn't make a jammer that was stronger than that."

As well as improving reception, the same technique can also be turned around so that a large number of weak radio transmitters, like the mobile-phone-level radios on small drones, can be combined into a powerful jammer.

İbrahim Kocaman, another student at the Naval Postgraduate School on Monterey, looked at how effectively a swarm of small drones might jam anti-aircraft radar. Normally this task is carried out by manned aircraft at longer ranges using specialized equipment in the form of jamming pods packed with electric wizardry. For example, the AN/ALQ-131 jamming pod can be carried by an aircraft in place of a bomb. It is about ten feet long and weighs over five hundred pounds, with a broadcast power of several thousand watts. Its range is classified and depends to a large degree on the radar in question, but is of the order of tens of miles.

By contrast, Kocaman was looking at drones with transmitters with just a tenth of a watt of power – that's the same broadcast strength as a cell phone – for his jamming swarm. Ten of them working together have less than a thousandth the power of the AN/ALQ-131.

Clearly, the swarm was not going to be able to jam radar at anything like as long a range, but it does not have to. As with the study on attacking an Aegis destroyer, the small drones are naturally small and stealthy, so they can get close to an enemy radar before they are detected.

In addition, Kocaman's paper demonstrates how ten drones could act together and use "distributed beamforming," like Kitts' formation robots, to achieve more than a simple addition of their power would suggest. Kocaman calculates ten of these drones could jam a Russian Sam-2 Fan Song anti-aircraft radar, producing too much radio noise for the radar to "burn through." This would render it helpless – the drones could broadcast the radar's exact location with impunity, and even "lase" it with a designator, making the radar a sitting duck for an air strike.

Kocaman estimates that a small, single-use jamming drone would probably cost no more than about $2,000 to build.

The effectiveness of swarm jamming is particularly significant when taken with the Aegis report, as Aegis is highly dependent on radar to spot incoming aircraft. A swarm that combined jammers with "flying IED" would be able to overwhelm a ship comparatively easily. If it attacked at night, even the last-ditch machine-guns would be lucky to score any hits.

Again, in Kocaman's calculation, the more drones that are added, the better the jamming – ten are vastly more effective than five. Twenty or fifty would be more effective still. Kocaman notes that even if several of the swarming drones were downed, the remainder still comprise an effective jamming force. Unlike manned aircraft, drones can tolerate dangerous missions with high casualties.

Kocaman also points out that the beamforming swarm is not limited to jamming. Its ability to act as a giant antenna means it could carry out stealthy electronic eavesdropping over a wide area. On a more humanitarian note it would be useful in search-and-rescue operations, as it could home in on a radio distress beacon from long range.

Invisible Sentinels

This idea of large numbers of drones joining together to work cooperatively may sound like some theoretical projection to the far future. But there are already teams of electronic sentinels in operation that use exactly the same sort of communication network that a drone swarm would need. This is the world of unattended ground sensors (UGS), sentries that quietly guard borders and perimeters. UGS have a lot in common with small drones. The major difference is of course that drones are mobile, but there are some striking similarities.

UGS go back to the Vietnam War, when early models were dropped from aircraft to detect passing Viet Cong supply convoys in an operation codenamed Igloo White. The sensors weighed twenty-five pounds and were the size of a fencepost – they were designed to stick into the ground on landing, with an operating life of about thirty days, a limitation imposed by the batteries of the day. Their radios could transmit a short range, so the only way to pick up information from them was via a specially-equipped aircraft circling overhead. And, needless to say, the early UGS were not cheap, but they did provide useful information.

Fast-forward to the modern age and, as you might expect, Moore's Law has effected quite a transformation. One example of a modern UGS is Lockheed Martin's Self Powered Adhoc Network or SPAN. This is a set of covert ground sensors far less conspicuous than the old version, and which require so little power they can operate indefinitely. Each of the SPAN sensor nodes is a palm-sized unit, usually concealed in a fake 3-D-printed rock.

They communicate via a mesh network; each node in the network communicates with all the adjacent nodes, so each acts as a router. The mesh is self-organising and self-healing, able to rearrange itself if some elements are lost.

The network also includes one or more dedicated gateway units. Each node only needs to connect to the next node, and with one of them in touch with the gateway, the entire array can send back information. The gateway can send back data to local ground troops via Wi-Fi, or it can use UHF radio or even satellite communication. A SPAN network located on the Afghan border can send information right back to headquarters in the US.

SPAN has a plug-and-play design, which means that new sensors can be added as needed. At present, the most likely sensors to be used are seismic units to detect ground vibration caused by passing vehicles or people, or acoustic sensors that pick up sound. Other variants include magnetic sensors to detect metal objects like passing vehicles, and chemical "sniffers."

Like small drones, SPAN has benefited from the easy availability of cheap computing power, which has reached the point where the sensors can carry out their own data processing. Their discrimination is good enough not just to tell that a vehicle has passed, but to tell what type, as they can even identify individual engines. Several sensors working together can triangulate the exact location of a particular signal. Rather than transmitting a lot of raw data for interpretation, they can send the summary, such as "group of several walking humans, fifty meters away to the northwest."

SPAN can run for about six months on battery power alone, but each sensor and gateway unit is equipped with a thin-film solar cell. Even beneath the rock camouflage, this provides enough power to run the device and recharge the battery so it can run indefinitely. Lockheed Martin has already sold thousands of SPAN units to domestic and foreign customers. The next generation is already on the way, which will have an even smaller electronic core and will be even cheaper. The current sensors are around $1,000 each, but future sensors could be so cheap they are scattered around freely to cover ever wider areas in greater detail.

Unmanned ground sensors are good for providing 24/7 surveillance of remote areas with minimal manpower. Their capabilities could be extended further by teaming them with small drones. Lockheed Martin is looking at an arrangement where a small drone, or a number of them, is situated near to SPAN networks. When an intruder is detected, SPAN could activate the drone, which would fly over and get "eyes on" with cameras and other sensors for a closer and more detailed look.

The system would not need to call a human operator until it had confirmed that there was a real intruder and not a false alarm. And if the small drone was of the lethal variety – such as Lockheed Martin's own Terminator drone described in Chapter Three – the operator would do little more than press a button for it to home in and deal with the intruder.

Combine and Fly As One

There are other ways that networks like SPAN could be integrated with small drones to make them even more effective. The European AWARE project – "Platform for Autonomous self-deploying and operation of Wireless sensor-actuator networks cooperating with AeRial objEcts" - combines a network of ground sensors with small drone helicopters.

The idea is that the helicopters can pick up the sensors and relocate them as needed. This takes air delivery of ground sensors to a new level: not only can the entire network position itself as precisely as needed in an area of interest with no human involvement, it can also move around as needed. If some of the sensors are lost, or if there is too much traffic to monitor in one particular area, it can rearrange itself to get a better look. And once the mission is over, AWARE can activate its helicopters again to return the sensors to base and do everything except pack themselves away in crates for next time.

The AWARE project has also explored a different sort of cooperation between drones: having multiple helicopters lifting a heavy load between them. This is not feasible (or at least not safe) with manned helicopters. After some hair-raising experiments in the US and Britain, the technique was abandoned. Helicopters are tricky enough at the best of times, and having two of more flying close together with an unstable load proved too dangerous to be worthwhile. Drones are different though. They react more quickly and can easily be networked together into a single entity. The AWARE team believes that it should be possible to get eight to lift a load cooperatively.

The GRASP (General Robotics, Automation, Sensing, and Perception) Laboratory at the University of Pennsylvania has also developed cooperation between small rotorcraft for carrying loads between them using similar principles.

Researchers at ETH Zurich have demonstrated swarms of small hexagonal drones that can link themselves together into flying rafts. What's interesting about this setup is that, as with other swarms, there is no one drone in control. Each unit decides what actions to take to keep the group in the air in what's known as a "distributed flight array," which works in like insects cooperating to carry a heavy load. The rafts can, in theory, be added together to produce a raft of any size. With enough drones, you could carry a load of any weight.

Group Brain

Perhaps one of the most intriguing possibilities for the swarm is that rather than just combining the power of their cameras or their lifting power, they merge their processing power into a group mind – or more accurately, what is known as a computer cluster.

Professor Owen Holland at the University of Essex became fascinated with insect behavior after spending time with an ant biologist trying to understand ant behavior. He found that seemingly complex behavior could be replicated in robots with just a few simple rules. He went on to develop what he terms a Gridswarm.

"It's what nature could do if we had telepathy," Holland told me in an echo of Selous' "thought transference or what?"

The Gridswarm is comprised of small drones that network together to form a single computing unit. Its power depends on the number of units available, and it can be a flying supercomputer with enough members. His initial version was more modest though and was comprised of several toy helicopters, each equipped with a Gumstix computer the size of a piece of chewing gum with Bluetooth communication.

Holland suggested that by combining their computing power over the Bluetooth link, the drones can take on challenging tasks such as mapping a cloud of pollution in three dimensions in real time and tracking it back to its source. Of course, this type of group intelligence would be equally applicable to military tasks such as identifying targets. The larger the swarm, the greater its processing power and the faster and more effectively it would be able to process the vast amounts of data it generates.

Kill CODE

Swarms are starting to move out of the laboratory and towards the battlefield. Perhaps the most significant development in this area is DARPA's CODE project – "Collaborative Engagement in Denied Environments." The aim is to develop software that will turn existing small drones into elements of a larger swarm.

DARPA's developers list some familiar potential advantages, including increased range through formation flying, increased area coverage, having a mix of different sensors (different drones might have radar, cameras, LADAR or other sensors), and simultaneous attack with overwhelming numbers.

For the post-attack phase, the drones would carry out real-time battle damage assessment and dynamic target reassignment. In other words, as soon as one target is destroyed, the rest of the swarm switches to the next target, so there is neither underkill nor overkill.

CODE has two main areas of development: vehicle autonomy, giving each drone the minimum level of intelligence to act on its own, and the team-level autonomy of the swarm. Vehicle autonomy includes automated take-off and landing (which have been such a bugbear with unmanned aircraft in the past), navigation, and dynamic flight planning. The team autonomy will concentrate on ways of fusing data together from the members of the swarm into a single picture, always a challenge with limited bandwidth. Once the picture is assembled, new algorithms will detect and identify targets more quickly and reliably, enabling a faster reaction and what DARPA bluntly terms a "better kill ratio." Many DARPA projects never come to fruition, but if successful, this one could help change perceptions of the power of swarms.

In a picture that is becoming familiar from other small drone research, one of DARPA's requirements is an open architecture with no proprietary restrictions. The agency has latched on to the benefits of the open-source approach and wants to build up a library of software modules that can be shared across different platforms.

Hide-and-Seek

Others are looking directly at the tactical implications of swarming drones and what they can do. One is the "multi-robot pursuit system," a software-based approach that aims to find the optimum searching pattern for a robot team in a cluttered urban environment while hunting for humans. This starts off relatively easy in a known environment but becomes trickier in unknown environments, including inside buildings. The robots have to understand lines of sight and appreciate which areas have been seen and which still need to be explored.

(As a curious sidelight, Lockheed Martin carried out a program on "covert robotics" developing software and sensors so robots can move around without being spotted by humans. The idea is that they can use their superior senses to detect approaching people and get out of the way. The project rapidly disappeared from public view, appropriately enough, and no doubt any continuing work is highly classified.)

Perhaps the most advanced game of hide-and-seek for robot swarms is the Urban Target Tracking program being carried out by UtopiaCompression Corporation for the US Air Force. This is developing a controller that does not just find people in a complex environment, it gets progressively better at finding them as it goes on. It uses a combination of machine learning and what is known as Monte Carlo simulation -- a mathematical technique based on probability, running the same process thousands of times and seeing what the outcomes are. These allow it to improve its ability to predict the movement of human targets. And not only does it track targets, it also observes and classifies behavior. It might, for example, be able to identify potentially hostile actions and distinguish insurgents from civilians.

The original Urban Target Tracking System used multiple drones to track one individual, and this is now being extended to tracking multiple individuals with changing environments and changing mission objectives. The company suggests that the system might also have commercial applications in border security and search and rescue. UtopiaCompression mentions the possibility of early deployment on the RQ-11 Raven. Swarming behavior is not just something for future drones; like other software, it can be retrofitted to existing machines.

The project provides a glimpse of what a future swarm might be able to do. Equipped with technology to extend mission time indefinitely – power line scavenging, perching, solar cells – the small drones could occupy an area continuously. They could track human targets and monitor their behavior with minimal human intervention. And with armed drones right on the spot, the sensor-to-shooter time is likely to be measured in seconds.

Not all swarm software comes from well-funded military programs. Plexidrone, funded via Indiegogo, is a small quadrotor for shooting video from a Toronto startup. It comes with SwarmTech software which allows up to thirty of the drones to fly together, controlled by one operator, capturing a scene from many different angles at once. The drones should start shipping in 2016. Numerous apps to take advantage of the swarming capability are likely to follow.

The Swarms Are Coming

As we have seen, a swarm is more powerful and harder to stop than a single drone and has many advantages over multiple drones acting independently. With the aid of a few simple rules, the members of the swarm can fuse together into a unit that can absorb huge amounts of damage and still carry out its mission.

Swarms can search rapidly and efficiently using the same sort of techniques as foraging ants. They can cover a wide area. And they are very good at hide-and-seek. They can help each other by flying in formation or combine their signals together for communications or electronic warfare. When they attack, their large numbers mean they overwhelm the target rapidly.

Swarms in nature tend to be homogenous – all the starlings in a flock are the same. There are some exceptions, though, such as ants that may have specialized workers and soldiers, or may even have multiple types of soldier and workers for different tasks. All the drones in a swarm may be the same, or the swarm may be composed of a mixture of different types – like SPAN, which can contain a mix of different sensors and communication nodes. Drones might simply be "hunters" equipped with more sensors and batteries or cheap "killers" with warheads. They might also include many types of specialists – with through-the-wall radar, or mapping LADAR or other sensors, or different types of weapon. The composition of a swarm would depend on its mission, so the swarm sent to search for insurgents in the hills would be different to one tasked with knocking out an oil refinery.

They may be hard to kill, but the question remains over how effective swarms can be at causing damage. A small drone might be able to knock out something light like a radar antenna, kill a person, but does the real business of warfare require heavier weapons?

We will see in the next chapter how terrifyingly lethal a swarm can be, thanks to recent advances in explosives technology.

REFERENCES

Chapter Swarm

1) "The morale of military units is prone to crack on the battlefield when casualties reach 25-35%."

Wainstain, L. (n.d.). (n.d.). The relationship of battle damage to unit combat performance. (http://1.usa.gov/1HIYzzg)

2) "A sobering assessment of what would happen if a US Navy destroyer was attacked by a number of cheap, unsophisticated drones."

Pham, L. (n.d.). UAV swarm attack: Protection system alternatives for destroyers. (http://bit.ly/21z8UEn)

3) A 2014 study by RAND analyses the effectiveness of different types of unmanned aircraft on a "hunter-killer."

Menthe, L. (2014). The effectiveness of remotely piloted aircraft in a permissive hunter-killer scenario. (http://bit.ly/1NJWup8)

4) "In 1986 Reynolds showed that movement of groups of creatures in nature, from schools of fish to flocks of birds or swarms of insects, can be modelled."

Reynolds, C. (1986). Flocks, herds and schools: A distributed behavioral model. (http://bit.ly/1O65san)

5) Termite mound building

Miller, P. (2011). *Smart Swarm: Using animal behaviour to organise our world*. London: Collins

6) Onyx swarming parafoil

Onyx autonomously guided parachute system. (n.d.). (http://bit.ly/1TBDQP9)

7) Swarm software

Axon AI. (n.d.). (http://bit.ly/1PB8rOm)

8) LOCUST

Smalley, D. (2015). LOCUST: Autonomous, swarming UAVs fly into the future. (http://1.usa.gov/1bez0GI)

9) "Precise formation control of large numbers of UAVs" – Lee Mastroianni interview with the author, June 2015

10) Wolf Pack Swarm Tactics

Coppinger, R. (n.d.). Wolf-pack (Canis lupus) hunting strategies emerge from simple rules in computational simulations. *Behavioural Processes*. (http://bit.ly/1RqgLQg)

10) Harris Hawk Tactics

Bednarz, J. (n.d.). Cooperative hunting Harris' Hawks. *Science.* (http://bit.ly/1OAAps6)

11) CARACaS (Control Architecture for Robotic Agent Command and Sensing)

Smalley, D. (2014). Navy's autonomous swarmboats can overwhelm adversaries. *ONR News.* (http://1.usa.gov/1m00c1z)

12) Benefits of formation flight for drag reduction

Blake, W. (n.d.). Drag reduction from formation flight. *US Air Force Research Directorate.* (http://bit.ly/1m00ghM)

13) "Christopher Kitts of Santa Clara University has been testing this approach as a way of boosting radio."

Hambling, D. (n.d.). Team-working robots huddle together to boost comms. *New Scientist.* (http://bit.ly/1XCXobD)

14) "We could have nulled a stronger jammer."

Garret Okamoto in interview with author, July 2010

15) "Looked at how effectively a swarm of small drones might jam anti-aircraft radar."

Kocaman, I. (n.d.). Distributed beamforming in a swarm UAV network. (Naval Postgraduate School thesis). (http://bit.ly/1XCXtfu)

16) Self-Powered Adhoc Network

Hambling, D. (2014). Network of 'Spies' scans no-man's land indefinitely. *WIRED*. (http://bit.ly/1NsAQiF)

17) AWARE project

Ollero, A. (n.d.). Platform for autonomous self-deploying and operation of wireless sensor-actuator networks cooperating with aerial objects. (http://bit.ly/1HIZpvW)

18) Distributed Flight Array (n.d.). (http://bit.ly/1NsAZml)

19) Gridswarm

Holland, O. (n.d.). University of Essex research home page. (http://bit.ly/1QWUN8i)

20) "It's what nature could do if we had telepathy" – Owen Holland interview with author, August 2005

21) Kill CODE

Establishing the CODE for unmanned aircraft to fly as collaborative teams. (2015, January 21). (http://bit.ly/1lct2vI)

22) Multi-Robot Pursuit System (n.d.). (http://bit.ly/1QWUVo4)

23) Covert Robotics

Hambling, D. (n.d.). Surveillance robots know when to hide. *New Scientist*. (http://bit.ly/1Q3y8H2)

24) Urban Target Tracking program

Intelligent cooperative control for urban target tracking with UAVs. (n.d.). Retrieved from Intelligent Cooperative Control for Urban Target Tracking with UAVs

CHAPTER EIGHT

MINIATURE TERMINATORS: THE POWER OF SMALL WEAPONS

> *"Most human problems can be solved by an appropriate charge of high explosives."*
>
> - Michael Tolkin, Uncommon Valor (screenplay)

Small drones do not look dangerous. To military professionals used to big bombs, they look like toys.

Although Switchblade drones have been effective against high-value targets in Afghanistan, they are gnats next to a modern combat aircraft like the F-35 Lighting II. The manned aircraft weighs thirty tons and roars overhead at several hundred miles an hour. When it attacks, several tons of high explosive rock the earth, sending up gigantic plumes of smoke. The Switchblade delivers the equivalent of a hand grenade, and a buzzing thing little larger than a pigeon does not look like a substitute for the raw power embodied by the F-35.

In terms of weight, even a whole a swarm of small drones is puny compared to the striking force of a manned plane. A single F-35 can deliver over fifteen thousand pounds of munitions in one sortie. A Switchblade-sized small drone delivers less than a pound. Even if several thousand drones were coordinated against a target, thousands of tiny pinpricks hardly seem to compare with being stabbed once with a full-sized spear. Big planes dropping big bombs is what strategic air power is all about.

Bombs today are comparable to those used in WWII in terms of size and explosive power. One of the recurrent themes of this book is how things can get smaller and more powerful at the same time. We have consistently seen electronics getting smaller: computers shrink from room-sized to pocket-sized; bulky, analog cameras are replaced by tiny digital ones; GPS shrinks from backpack to microchip. Analog objects can be compressed with digital technology: a shelf-load of books, vinyl records, or movie reels now fits in the palm of your hand.

Some things, though, defy the best efforts of scientists to make them smaller. Meals are the same size as fifty years ago, despite efforts by food technologists to reduce lunch to pill form. The hammer you drive nails with may be space-age alloy, but it is no smaller than the one in your grandfather's toolbox.

In this respect, bombs look more like hammers than computers. Intuitively, we feel that if you're looking for more bang, you want something big, and that there is no substitute for weight of explosives. Bombs are described in terms of pounds of explosive, and nuclear weapons are rated by their effective equivalent in thousands or millions of tons on TNT.

In reality though, the tiny drones are not a change of direction, but the continuation of a trend that has seen bombers doing more with less for several decades. Small and smart is more deadly than big and dumb, a lesson that goes back as least as far as the tale of David and Goliath. A well-placed pebble beats a Goliath's giant spear with a twenty-pound iron head every time. Small drones are the smallest and smartest weapon yet, and that makes them more powerful than anything that has gone before.

Carrier Power

To put this in context, we need to look back to the start of the twentieth century. In those days, the battleship reigned supreme. These were armored behemoths, brandishing guns that could rain down one-ton shells on a target twenty miles away. They were virtually immune to any but the heaviest artillery. The term *gunboat diplomacy* was coined to describe the policy of sending in warships to threaten minor nations that might cause trouble.

The Anglo-Zanzibar War of 1896 is often described as the shortest recorded war in history. A new Sultan was appointed without the permission of the British consul, and this was considered an act of rebellion. A Royal Navy squadron was on hand to deal with the revolt. The new Sultan took refuge in his palace, protected by his guards, artillery, and machine guns.

A short exchange of fire followed.

After thirty-eight minutes of bombardment from British vessels, five hundred people had been killed or wounded in and around the palace, and the Sultan surrendered. One British sailor was injured in the action. The brief war was an object lesson in how less advanced nations had no answer to the superior firepower of the armored warship. The big ships did not just mean major powers fought on advantageous terms; they made any fight so one-sided as to be farcical.

The successor to the battleship is the aircraft carrier, a mobile platform with its own air force capable of sweeping away any but the strongest opponents. The aircraft carrier battle group or carrier strike group is the embodiment of long-range force projection. When you want to express commitment or apply military pressure to a remote part of the world, a carrier strike group is the deluxe option with all the extras.

Russia has one operational carrier; China has one and is building another. The French have one, but it will be out of service from 2015-18. Britain is building two, but only one will go into service and will not be fully operational until 2020. The US has ten carrier groups.

If there is a situation brewing in the Gulf of Sirte or the Indian Ocean, the US will send a carrier group. It is not quite gunboat diplomacy, but it underlines the point that the US can and will intervene wherever necessary. The Air Force of Tanzania, the modern-day state of which Zanzibar is a part, has a handful of old MiGs, which would be no match for a carrier air wing. A second encounter between the Africans and Western naval power would go the same way in about the same time.

A US Navy group comprises one aircraft carrier, one cruiser, two or more destroyers and various other support ships including submarines, minesweepers, supply vessels, and other auxiliaries. The rest of the ships are just there to protect the aircraft carrier from various threats and ensure it can carry out its mission. This is what the modern Navy is about; according to one survey almost half of Navy personnel serve either on or directly supporting carriers.

The biggest and most expensive ship in the inventory, an aircraft carrier is a truly impressive piece of military hardware. Each one takes five years to build. You can get some idea of just how impressive they are on a tour of the USS Midway, a carrier commissioned in 1945 that served until 1992 and is now a visitor attraction in San Diego. The flight deck stretches out over four acres, and it feels like an extension of the city. Carriers are sometimes described as floating towns, and it is an accurate description. The Midway and her sister ships each had a sailing crew of over three thousand and hosted an air wing with over two thousand personnel maintaining and flying the aircraft. This travelling population brings requirements for accommodation, feeding, and laundry, not to mention dental and medical facilities, shops, and water distillation plants supplying hundreds of thousands of gallons of fresh water a day.

This mobile township has its own power plant, or rather twelve of them, driving four turbines that can send the whole sixty-thousand-ton mass barreling through the water at thirty-five miles an hour.

This entire floating community of over five thousand people exists in order to keep the aircraft flying. When a major attack is called for, a carrier can launch an Alpha Strike comprising about half its total complement of aircraft. An Alpha Strike is a "deckload" of planes, the greatest number that can be brought to the flight deck, armed, fueled, and launched in a coordinated mission. The rest of the aircraft will be getting ready or returning from other missions.

During the Vietnam War, the USS Midway usually carried sixty-seven aircraft, a mixture of F-4 Phantom fighters, A-4C Skyhawk attack aircraft, and RF-8A Crusader reconnaissance planes along with A-3B Skywarrior electronic warfare aircraft. When she was tasked with carrying out an attack, the Alpha Strike package would include perhaps twenty-four aircraft carrying bombs, plus their escorts.

Twenty-four aircraft, each with a dozen 750-pound bombs – over a hundred tons of bombs in total – sounds like enough to devastate anything. In reality, dropping bombs in the general area of the targets is not enough. Precision is the key.

During the Vietnam War the average distance by which a bomb missed its aim point, known as "circular error probable," was about four hundred feet for bombing from medium altitude. Low-altitude bombing was more accurate, sometimes much more accurate, but was too risky. Four hundred feet may sound haphazard, but computing bombsights had improved greatly since WWII, when the average miss distance was three thousand feet. This is why bombing in WWII focused on large, spread-out targets like cities: bombers were not accurate enough to hit compact, military targets. When the objective was to hit something as specific as a factory, planners had to send large numbers of aircraft to score a single hit.

The real difficulty comes when attacking something like a bridge, where the target area of a roadway or support beam is just a few feet across. An entire Alpha Strike's twenty-four planes may not be enough to score one solid hit.

The best-known example of this problem came with the strategically vital Thanh Hoa Bridge in the Vietnam War. The bridge was the frequent target of Alpha Strikes from Navy carrier Air Wings. The entire deckload of aircraft would release their bombs without doing enough damage to put the bridge out of action. Traffic was interrupted on several occasions, but because the bridge supports remained intact, any damage was easily repaired. Eleven aircraft were lost during these attempts.

The bridge was eventually knocked out using new technology: "smart bombs" with laser guidance. This is now familiar from laser-guided weapons like the Hellfire: instead of falling where it is dropped, a seeker head on the bomb locked steers it towards a laser spot projected on to the target. If everything works – if the bomb is released on the right trajectory, and the seeker head sees the laser dot, and the operator keeps the target illuminated – the bomb hits within a few feet of the aim point.

In 1972, eight F-4 Phantoms equipped with new PAVE KNIFE laser designators and two-thousand-pound laser-guided bombs attacked the Thanh Hoa Bridge. A single guided bomb directed by the new designator squarely striking one of the bridge's supports was enough to bring down a span. The Navy sent in follow-up strikes to make sure, distrustful perhaps that so hard a nut could be cracked so easily. But that one strike should have been enough.

The success of laser-guided weapons at Thanh Hoa was the start of a revolution in "precision guided munitions." Old "dumb bombs" were fitted with laser guidance kits to become smart bombs. A small number of guided bombs could do the job of hundreds or thousands of unguided bombs, striking targets that would be impossible by conventional means.

Although their accuracy was overstated at the time, the 1991 Gulf War was notable for daily video presentations of smart bomb strikes, showing the weapons hitting bunkers, bridges, and individual tanks and other vehicles with seemingly unerring accuracy. The Paveway series of guided bombs that are currently in use are direct descendants of the PAVE KNIFE system used in Vietnam.

According to a popular analysis of this improved accuracy, one modern laser-guided bomb was as effective against a point target as thirty Vietnam-era F-4 Phantoms dropping their entire load of bombs. In other words, one precision-guided weapon is better than a whole Alpha Strike package using unguided weapons. (Using WWII technology, it would take the bombs from an incredible fifteen hundred B-17 Flying Fortresses to hit the same target.)

During WWII there strategic bombing was directed at "choke points" in German industry. The ball-bearing factories in Schweinfurt were identified as being key to German manufacturing of tanks and aircraft. However, even massed bombing raids by hundreds of aircraft were not effective because too few bombs hit the actual target.

Precision means that hundreds of bombs are not wasted on the area around the factory. They can all hit the factory itself. A similar operation today would only require a handful of aircraft. Smart bombs though are not truly smart. All they do is go exactly where they are told. Unlike small drones, they cannot send back information about the target and get a close-up view. Weaponized drones are smarter than smart bombs.

Weapons like Switchblade allow the operator to see exactly what they are hitting in the factory. They can avoid the canteen, the store rooms, and the less significant parts of the production lines. Two key pieces of machinery are involved in the process of call-bearing manufacture: the "cold header," which smashes lengths of thick wire into rough spheres under huge pressure, and a "flash machining" instrument to shape the ball bearing into its final form. Destroy either of these two expensive, complex machines and the entire factory grinds to a halt. Being numerous and able to scout up close, small drones could locate and identify specific machines. The operator could then put them out of action with special warheads such as thermite (see below). A handful of drones could succeed where hundreds of heavy bombers failed.

Sometimes the target is simply a single important person, in which case a guided weapon precise enough to hit them and powerful enough to kill them meets the exact requirements.

Numerous aircraft were dispatched against individuals such as Saddam Hussein, with little success. On one occasion, the 2,000-pound bombs intended for a restaurant where the Iraqi leader was eating apparently demolished family homes nearby, with many casualties. A few perching drones might have been more effective, able to not only confirm the target from close up but also to guarantee a kill. Similarly, the mission to kill Osama bin Laden might have been carried out by a handful of small drones. It would not have the same heroic swagger as Seal Team Six's operation, but it would have been far less risky.

A swarm of a thousand Switchblade-type drones could in theory kill a thousand targets dispersed over a wide area. This is far more than any bomb or missile of similar size. What's more, they could (in theory) confine their attacks to the armed individuals in an area. It is a revolutionary capability, and one that an F-35 could never achieve – unless it was acting as the mothership or delivery platform for a drone swarm.

A Smaller, Deadlier Bang

Precision guidance means that fewer bombs are needed. Rather than a hundred planes, you only need one. The next question is how big a warhead it takes to attack strategic targets like bridges and buildings.

Two thousand pounds has long been the standard bomb size deemed necessary for destroying structures made of reinforced concrete. It is used today just as it was forty years ago. According to conventional wisdom, anything less than a two-thousand-pound bomb, containing about a thousand pounds of actual explosive, will leave a shallow crater without doing critical damage to a concrete target.

Piling on enough explosive to do the job is a brute force technique. As with getting the explosives to the target, intelligent application means less effort. If you lay explosives by hand, about two hundred pounds of well-placed C4 will do the job of a two-thousand pound bomb. Two hundred pounds is a lot to carry, so US Special Forces have developed a lightweight concrete-busting tool.

The M150 Penetrating Augmented Munition (PAM) is a portable demolition device weighing just forty-two pounds. First introduced in 1998, it is highly effective against reinforced concrete structures. When triggered it goes through a complex, four-stage firing process that is like a speeded-up version of the miner's approach of drilling and blasting.

PAM's first charge punches a tunnel deep into the target. The subsequent stages cut through any steel reinforcing bars, propel a powerful explosive charge into the tunnel, and detonate it. Concrete is strong in compression, but weak in tension. It is almost impossible to crush a concrete block, but comparatively easy to tear it apart from inside. That's how PAM can replace a two-hundred pound charge of C-4, or the warhead on a 2,000-pound bomb, and demolish a reinforced concrete structure such as a bridge support measuring fifteen feet by five by six. It would take a team of seven about three hours to rig a target with C4 explosives for demolition, whereas with PAM, the same process takes about two man-minutes

In theory, one PAM could have disabled the Thanh Hoa Bridge. An explosive charge correctly placed by one commando could do what the total striking power of a carrier like the USS Midway with her crew of thousands – backed by thousands more on its escort and supply ships – failed to do with dumb bombs. The precise application of force is more effective than blind pummeling. And small drones excel at hitting a precise spot.

At forty pounds, something like PAM is too heavy for a single Raven-sized drone. But as technology improves, demolition charges are getting smaller still.

Reactive Materials

Rather than talking about explosives, researchers tend to refer to "energetic materials." It is an umbrella term covering everything from detonators to incendiaries to rocket fuel. Energy may be produced in the form of heat, or shockwaves, or accelerating fragments to high speed ("driving metal"), and researchers attempt to tailor materials to the exact requirements of a particular mission.

In the field of energetic materials, Reactive Materials or RMs have shown great potential for developing weapons far more effective than conventional high explosives. RMs typically consist of a material such as Teflon mixed with metal powder.

Reactive materials also make highly effective shrapnel. Normally, shrapnel is made of steel or similar material; shrapnel fragments are like miniature bullets. But reactive material shrapnel is explosive: the material can be engineered so that it starts releasing energy when it impacts an object. This makes RMs effective as anti-aircraft and antimissile warheads, as adding a little explosive power makes them much more lethal. According to one estimate, they are five times as effective against aircraft and similar targets as conventional shrapnel. They would be similarly effective as an anti-personnel weapon.

During WWII, a new type of weapon was developed known as thermite. This is a simple mixture of metal and metal oxide powder, like iron oxide and aluminum, but it burns at extremely high temperature. Impossible to extinguish once started, thermite can melt through steel plate, and commandoes used thermite charges to disable guns and heavy machinery. Reactive materials can do better than that.

Energetic Materials & Products Inc. of Round Rock, Texas, has been involved in the Air Forces' micro-scale ordnance efforts and used the technology in a spin-off called the Tec Torch or Metal

Vapor Torch. This flashlight-sized device blasts out a flame jet that cuts through metal like a hot knife through butter, slicing through a half-inch steel bar in less than a second. It has been designed as a breaching tool for police and others who need to cut through bolts, chains, and padlocks at high speed. The Tec Torch is based on reactive material technology with solid fuel and oxidizer, and is cheaper, lighter, and more compact than the traditional oxyacetylene cutting torch.

Each fuel cartridge weighs a couple of ounces and contains precisely graded particles of magnesium, aluminum, and copper oxide. This resulting flame jet burns at over three thousand degrees centigrade and has a speed of over two thousand meters a second. A rectangular carbon fiber nozzle shapes the jet into a flat blade for cutting through bars. The jet has higher energy density than a gas flame, and the cutting action is a combination of heat and abrasion by particles of metal oxide.

The Tec Torch gives an indication of the sort of weapons that may be available for small drones. A drone perching on a structure could use its own version of the Tec Torch to slice through a vital component, such as power or communication lines – or the cables supporting a suspension bridge. One drone might not be able to cut a cable on its own, but several drones successively cutting at the same spot certainly could. A swarm of many drones could cut through many cables.

This type of technology could also be effective at puncturing pipelines, and fuel and chemical storage tanks.

Another application of reactive materials is in demolitions. The PAM described above is a complex, multi-stage device. In 2007, the US Army started work on a simpler demolition charge than PAM that blasted a jet of reactive material into a concrete target. The reactive material jet explodes as it strikes, tearing the concrete apart.

The Army's first project using this approach was called Barnie; this was scaled up into a forty-pound device called Bam-Bam. Bam-Bam was tested against various structures, including a full-size reinforced concrete block that was five by six by eighteen feet – like a bridge support – with satisfactory results.

ATK, a company with some of the most advanced developments in this area, has continued the work since then. In one test a charge of less than twelve pounds produced a crater eight feet across and more than four feet deep in a concrete target. Little information has been published in this field in the last few years, but the latest generation of RM-based weapons for destroying concrete are likely to be smaller and more effective than their predecessors.

A series of small charges can do the job of one large one. The Air Force refers to the technique of "laddering" – setting off several charges in the same spot in succession – to break through concrete too large for one charge. This suggests that a dozen small drones targeting the same point in succession could do the job of something like PAM, relentlessly chipping away until it failed.

However, in most cases it is not even desirable to have all the explosives in one spot. Demolition professionals prefer to use multiple, smaller charges.

Normally, structures are not brought down by huge charges like a 2,000-pound bomb, or even a forty-pound charge. The demolition of Cleveland's I-90 Innerbelt Bridge, a streel construction over a hundred feet tall and wide and just over a mile long, is a good example. In July 2014, the entire structure weighing some five thousand tons was brought crashing down by the simultaneous detonation of eighty-eight charges with a combined weight of just a hundred and eighty-two pounds, or a little over two pounds per charge. The concrete supporting piers, which would have been targeted in a military strike, were slated for later demolition. Without a roadway they are of little use. The Innerbelt Bridge was well and truly destroyed by less than two hundred pounds of explosive.

Tall buildings are generally demolished by an implosion in which key structural components are destroyed so that the building collapses in on itself without any damage to surrounding buildings. There are plenty of examples of this type of demolition on YouTube, including a thirteen-storey building in Coral Gables brought down with ninety pounds of explosives, and the three giant concrete cooling towers of Didcot power station, which took just over a hundred pounds each (in a large number of small changes) to reduce to rubble. In each case the demolition did not involve one big blast but a multiplicity of smaller explosions, carefully placed and timed for maximum effect.

The key with demolition in the civil sector is not simply piling on enough explosives to make sure the job is done. It is a matter of the highly skilled application of minimal amounts of explosive in vital spots. This is the type of demolition that a coordinated swarm of small drones could carry out, aided by the unique precision that working up close brings.

There have not as yet been any demonstrations of this type of attack, although USAF documents refer to experiments with cooperative, simultaneous attacks involving multiple small warheads. Given that the principle is already well established for manual-laid explosives, there is no specific reason to doubt that drones could do the job as well as explosives placed by hand.

Enhanced Blast: "Thermobaric Urban Destruction"

Smaller structures, such as houses, can be easily demolished by a new type of explosive known as "enhanced blast" weapons, or thermobarics, without the need for careful placement of charges.

Conventional explosives are composed of large molecules that break down and release energy. The military explosive RDX ("Research Department Explosive"), also known as cyclonite or hexogen, has a chemical formula of $C_3H_6N_6O_6$ and a complex branching structure based on a central ring of nitrogen atoms. Those bonds are unstable, and when they are broken, the explosive detonates with a velocity of more than eight thousand meters a second.

By contrast, thermobarics do not explode at all; technically, they just burn very fast. Some types have their own oxidizer, but some simply react with oxygen in the air. In its simplest form, enhanced blast can be achieved simply by adding finely powdered metal such as aluminum to an explosive charge. More sophisticated versions consist of nothing but powdered metal and oxidizer; the explosive is released into a cloud, which is then set off with devastating effects.

Thermobarics are typically several times as powerful as TNT by weight because the oxidation reaction is more energetic than the breakdown of an explosive molecule. However, what is more surprising is that thermobarics are so far more destructive than condensed explosives with the same power. This is because the blast from an expanding thermobaric fireball goes on for longer than a normal blast. It still only lasts a matter of milliseconds, but the increased duration makes it more effective at bringing down walls. Just as a low-pitched sound from a passing truck will cause windows to rattle, or a high-pitched note can shatter a wine glass, the blast from a thermobaric explosion is 'tuned' to cause more movement and more damage in building walls.

Researchers found that the blast effect of thermobarics was greater in enclosed spaces where there was turbulent mixing during the explosion and the thermobaric mixture burns efficiently.

Thermobarics have been well proven in battle for more than a decade. The SMAW ("Shoulder-Launched Multipurpose Assault Weapon") is a bazooka-like weapon used by the US Marines. An enhanced version for urban combat was used extensively in Afghanistan and Iraq, notably in the battles for Fallujah. Known as the SMAW-NE (for Novel Explosive) , the new warhead contains four pounds of a mixture known as PBXIH-135, which combines a standard plastic explosive (PBX – Plastic Bonded eXplosive) with a precisely calibrated amount of finely powdered aluminum.

Troops gave highly favorable reports of the effects right from the start, with one post-action report stating:

"One unit disintegrated a large one-story masonry type building with one round from 100 meters. They were extremely impressed."

The after-action reports from operations in Fallujah bore out the SMAW-NE's ability to bring down buildings:

"Where possible, they were placed on a roof looking down, where the SMAWs could smash three or four houses in advance of the squads."

Large windows reduced the effect of the blast inside buildings, but "SMAW gunners became expert at determining which wall to shoot to cause the roof to collapse and crush the insurgents fortified inside interior rooms," according to a piece in the Marine Corps Gazette. Over a thousand were fired in Fallujah alone.

This type of weapon gives infantry the sort of power usually reserved for heavy artillery. It also means that footsoldiers can cause damage on a massive scale. The action in Fallujah drew heavy criticism for the level of destruction involved. Such weapons are only appropriate when everyone in the area is a combatant.

One limitation was that the new SMAW round was far more effective inside a building than in the open air. Marines started using a two-stage approach: firing one of the old high-explosive SMAW rounds to make a hole in a wall, then firing a thermobaric round through the hole into the interior. This ensured that the second round "would incinerate the target or literally level the structure."

Makers Talley (later acquired by the Swedish/Finnish company Nammo) advertised the new version of the SMAW in a brochure titled "Thermobaric Urban Destruction." This included test pictures showing a two-storey building disintegrating into a pile of rubble with one hit. The brochure was withdrawn after a piece by this author drew attention to it.

The same type of explosive has been used in various other weapons, including one version of the Hellfire, the AGM-114N "metal augmented charge" in which the usual high-explosive warhead is enhanced by the addition of a jacket of aluminium powder.

There are smaller versions as well, including the XM1060 fired from a 40mm grenade launcher, used in Afghanistan in 2003. Reports from the field suggest that even a weapon this size is capable of collapsing mud-brick buildings and it has proven highly lethal, especially in caves and tunnels.

Part of this effectiveness is because the prolonged blast tends to flow around obstructions and travels for long distances indoors or underground. Thermobarics have become the weapon of choice for clearing caves and bunker complexes.

The prolonged blast is also more lethal against human beings than other types. You might survive a blast of forty pounds per square inch from a condensed explosive, but just ten pounds per square inch for a few milliseconds longer from a thermobaric blast will pulverize your lungs. Body armor and sandbags offer no protection from this sort of damage.

This type of warhead is well suited to small drones. Unlike the rocket fired by the SMAW, the small drone is a guided weapon and not limited to approaching a building from one side. It can fly around and find a suitable window or other entry point, potentially even arriving down chimneys. Anything short of a sealed bunker will have entry points.

Enhanced warheads have improved considerably since the early 2000s. Modern production techniques allow the exact size, number, and distribution of particles in the mix to be to be controlled precisely. In particular, the technology for producing nanoscale particles of aluminum, and storing them safely, has progressed

Manufacturers are reluctant to discuss details, but nano-aluminum is now of a much higher quality.

More important, in the early days engineers were guided more by trial and error. More recently they have been able to accurately model the complex physics of the thermobaric fireball with high-fidelity computer modelling. An ideal thermobaric round would be structured so that all the fuel burned effectively in the fireball. Researchers now know how mixing that occurs indoors can be created – and perhaps bettered – by structuring the explosive so that jets of particles are produced during the blast. Unclassified results from one Canadian research group suggest that it should be feasible to make warheads around five times as powerful as existing munitions without changing the ingredients.

A small drone with this type of warhead could undoubtedly bring down a two-storey house of normal construction. A sobering thought when such a drone could be sent anywhere in the world with precision of a few feet.

A swarm of ten thousand small drones could level a town.

Although the thousand pounds may be less weight than the bomb load of an aircraft, a thousand drones would do far more damage than any single aircraft. As before, their ability to pick out individual targets and strike at exactly the right spot makes them more effective than unguided brute force.

Twisted Firestarters

An arsonist can do tremendous damage with one lighted match, and incendiaries may be the weapon of choice where the payload is limited. Even a small fire can quickly spread to engulf a building, a city block, or a forest. This was how the Japanese hoped to inflict serious damage with the Fu-Go balloon bombs mentioned in Chapter 1.

The military have preferred to use incendiaries on a gigantic scale. In WWII in Europe, massed Allied bombers would attack first with high explosives to break open buildings, followed by a wave of incendiaries to start fires. In Japan the buildings were less solid, and Boeing B-29 Superfortresses carried out pure incendiary raids on Tokyo and other cities. They dropped the M-69, a hexagonal steel pipe three inches across and twenty inches long filled with a newly-invented jellied gasoline mixed with phosphorus known as napalm. The pipe was heavy enough to break through roof tiles and penetrate into the rooms below; a few seconds after impact, the M-69 threw out flaming gobbets of napalm, which stuck to anything and burned whatever they touched.

"A kind of flaming dew that skittered along the roofs, setting fire to everything it splashed and spreading a wash of dancing flames everywhere," according to one eyewitness.

Thirty-eight M-69s were bundled together in a "cluster bomb" that split apart midair and scattered its contents over a wide area. Each B-29 carried forty clusters, making over fifteen hundred M-69s per aircraft.

The plan was to start so many fires at the same time that it would be impossible to extinguish them. It worked exactly as intended. The fires spread out of control; people fled, diving into rivers and canals and drowning, or huddling in shelters where they suffocated and burned. A single raid on Tokyo in 1945 burned down fifteen square miles of the city, destroying hundreds of thousands of homes and killing over eighty thousand people. Weather conditions were not right for a firestorm, a phenomenon in which the combined strength of the fires creates a powerful updraft and sucks in more air with hurricane-force winds. Firestorms had occurred in Dresden and elsewhere; one in Tokyo would have made the attack even more devastating.

"We scorched and boiled and baked to death more people in Tokyo on that night of March 9-10 than went up in vapor at Hiroshima and Nagasaki combined," claimed General Curtis LeMay. Although not quite accurate (the atomic bombs killed over 130,000, the Tokyo firebombing about 100,000), it shows how the atomic bomb was merely an extension of existing bombing.

The outcome of the Tokyo raids might have seemed satisfactory in terms of destruction, but the analysts noted that the vast majority of incendiary bombs were wasted. Bombing inaccuracy meant that many missed the target area entirely. Most that landed in the right area had no effect: they landed in roads, in gardens, or in parkland or other empty spaces. Even those that did hit roofs sometimes bounced off or lodged uselessly in chimneys.

This sort of attack would be more efficient if an incendiary could be placed inside a building. In the right place, even a tiny incendiary would be practically guaranteed to start a fire. One ounce of napalm could be more effective than a dozen M-69s scattered at random, just as one aimed bullet is more effective than a thousand sprayed aimlessly.

This led to one of the most bizarre plans of the war, which makes even the Fu-Go look ordinary. It all started when biologist Dr. Lytle Adams noted that the humble bat might be capable of carrying "a sufficient quantity of incendiary material to ignite a fire."

Project X-Ray involved capturing thousands of bats and putting them into a state of hibernation by refrigeration, taking advantage of the bats' natural tendency to sleep when the temperature drops. Each bat could then be fitted with a tiny bomb. The bats were packed into special trays which were in turn fitted into bomb casings, which would be dropped on Japanese cities. Released mid-air the bats would naturally seek refuge and roost in the eaves of houses – after which the incendiary bomb carried by each bat would burst into flames.

The researchers found that a half-ounce bat could carry a load weighing more than itself. A suitable incendiary device was devised, a celluloid capsule filled with napalm with an igniter the size of a match head. It worked in a similar fashion to the static line used by parachutists that automatically pulls the ripcord. In this case, as soon as the bat flew free from the bomb it pulled a pin, releasing a chemical that ate through a wire and triggered the napalm in fifteen minutes.

The bats would die in the process, but this was viewed as an acceptable sacrifice for the war effort.

Thousands of bats were successfully captured and released "unarmed" from aircraft, duly finding roosting places in buildings below. There were to be no tests with bats carrying live bombs, so there was no fire crew assigned to the project. Firefighters would add to the security risk on the secret project.

Disaster struck at Carlsbad Auxiliary Airfield in a test when the bats were not supposed to be released. The X-Ray team was filming the effects of the bat bomb indoors. Live bombs were attached to six hibernating bats. The cameramen were warned that it would only take the bats a few minutes to warm up and become active again. Unfortunately the cameramen did not realize just how active bats can be. Frantic efforts failed to net any of the six armed bats and they flew off, seeking places to roost.

At least one of the six headed for a new control tower, another for a newly-built and unoccupied barracks building. Exactly fifteen minutes after the bombs were armed, both structures burst into flames. The fire rapidly spread in the dry desert conditions, consuming hangars and offices. It was too late to save the airfield buildings, but not too late to maintain security. Baffled firefighters who arrived to tackle the blaze were turned back from the gates while the buildings continued to burn. A few days later the burned remains were bulldozed to hide the evidence.

Surprisingly enough, this fiasco did not result in the project's being cancelled. What finished it was the question of timing. The bat-bombs would not be ready for operational use until sometime in late 1945, by which time there would be virtually no targets left for them to attack. And, as it happened, there was a more powerful secret weapon for attacking the remaining Japanese cities that would be ready first.

A small perching drone could deliver multiple incendiaries the size of the bat bombs. There are already drones that can break windows (as we will see shortly), and a perching/walking drone could find its way inside and place the incendiary where it would be most effective. Unlike the wayward bats, the drones could be directed to exactly the right buildings. Also unlike the bats, the drone could leave an incendiary and fly away to film the results from a safe distance.

A swarm of hundreds or thousands of drones could start a huge number of fires simultaneously, certainly enough to overwhelm any city fire department. Again, their ability to place each bomb precisely makes even a small weight of incendiaries far more effective than tons of napalm dropped haphazardly from larger planes.

Tank Busters

Small drones are highly effective against so-called soft targets like civilian vehicles. There was a specific requirement for LMAMS that it should be able to take out cars and pickup trucks, which are popular with insurgent groups around the world. There are plenty of videos proving that a small warhead is more than adequate for this type of target. Armored military vehicles are another story. A hand grenade will do little damage to a vehicle protected by an inch of steel plate. But yet again high precision and intelligent targeting make an effective substitute for brute force.

In the 1991 Gulf War, laser-guided Mk 82 bombs weighing five hundred pounds were used for "tank plinking" attacks against individual Iraqi tanks. Unlike in previous wars when dozens of bombs were needed to guarantee a hit on such a small target, laser guidance meant that a pilot could score four kills with four bombs. The bombs were accurate enough, and a bomb of this size was overkill even against heavily-armored Russian-made T-72 battle tank.

In the 2003 war in Iraq, the Hellfire missile weighing a fifth as much proved just as efficient at destroying tanks. Laser guidance meant that every shot was likely to find its mark.

You need a large warhead to destroy a tank because of all the armor. The T-72 has frontal armor more than eighteen inches thick, and the Hellfire can punch through it. But tank armor is not distributed evenly. Most of it is at the front, because that's generally where the threat comes from. The tank is built as low as possible, and the armor is concentrated at the front to have the maximum protection between the crew and the enemy, usually other tanks or foot soldiers with missiles and rockets.

There have been decades of research into small weapons to go through armor, ever since the development of the first Bazooka in 1942. This type of weapon has a shaped-charge warhead which, like Barnie, turns the force of the explosive into a narrow, armor piercing jet. The AT-4 is the Marine's current version of the old Bazooka. It looks like its predecessor but is vastly more powerful. The AT4's warhead weighs just under a pound, and it is capable of penetrating an impressive fifteen inches of armor compared to three inches for the original bazooka. This is still not enough to take a T-72 head on – tank armor is specifically intended to defeat this sort of threat – but it means the soldier can tackle anything else on the battlefield. And if he can get a shot at the side, rear, or top of a tank, he can damage that too.

Unlike a foot soldier, a drone doesn't have to go head on. And as we have seen, it can be very precise.

From above, the T-72 is a much easier prospect. The large, flat surface of the top of the tank has comparatively thin armor; if it was as thick as the front, the tank would be too heavy to move. The top armor on the T-72 is around two inches thick, and there are spots where it is even weaker.

While a small charge can breach the armor, the damage it does – the "behind armor effect" – is limited. One soldier compares it to firing a bullet through a car – alarming for the people inside but not likely to cause real damage. The high-speed jet of metal will injure anyone it hits and may set off fuel or explosives, but in a vehicle the size of the T-72, most shots will do little harm. That happens when the shot placement is more or less random, as it is likely to be in battle using an unguided weapon like the AT-4, often at long range against a target that may be moving. In practice it usually takes multiple hits from this sort of weapon to stop a tank.

The situation is different when the same warhead can be delivered by a drone. Instead of an unguided warhead, you have one with extreme precision. Current guided weapons sense a target and tend to aim approximately at its center of mass. (A major exception is heat-seeking missiles, which home in on hot exhaust pipes). As we have seen, a small drone has enough computing power to do something much more sophisticated.

The Munitions Endgame Geometry for Optimal Lethality (MEGOL) program is a new approach to maximizing the effectiveness of small weapons using improved sensors, guidance, and information processing power. MEGOL was developed for the US Air Force by the Survice Engineering Company, who patented it in 2008. MEGOL has a "lethality database" with details of the various vulnerabilities for each potential target; this would, for example, include all the weak spots on a T-72. MEGOL adjusts the attack trajectory as a missile or other munition approaches the target to ensure it hits as close as possible to the aim point.

With something like MEGOL, a drone could pick the location for maximum damage. It could target an individual crew member or a fuel tank or an ammunition storage bin where a hit will set off secondary explosions and destroy the entire vehicle.

Not Either/Or but Both/And

The existing warheads on Switchblade are anti-personnel/vehicle charges that combine blast and fragmentation effects. The different types of warhead above add several new types – demolition charges, enhanced blast, incendiary, and armor-piercing shaped charge.

It is common for there to be several different versions of a weapon for different targets, such as the Hellfire variants described in Chapter 2 for use against tanks or bunkers. The latest Hellfire, the AGM-114R, manages to combine the effects of several types into one warhead: it has armor-piercing, shrapnel, and thermobaric blast. Modern warhead technology is going even further, with multi-mode warheads that can produce different effects depending on how they are triggered.

General Dynamics Advanced Warhead Technology includes warheads that combine enhanced blast with an ability to penetrate armor and destroy buildings. The trick is in how the warhead is triggered. A multimode warhead has several different detonators that are fired in different combinations to produce different effects. It may be triggered in a way that channels the force of the blast into an armor-piercing jet, or an alternative combination can throw out fragments of reactive material in all directions. Or the entire warhead may be set off so that the effect is entirely one of blast – or, by shaping the fireball differently, it may maximise the thermal output for the greatest incendiary effect.

There is an even more advanced type of warhead on the way that goes by the name of MAHEM, short for Magneto Hydrodynamic Explosive Munition. Originating with DARPA in the 1980s, this has been kept very quiet, but a trail of documents indicates it progress. This is a true smart warhead that uses electromagnetic effects to alter its output. It works in two stages: the blast itself is enhanced, with a powerful electric current being passed through the explosion fireball to increase its detonation velocity and the blast pressure. The second stage captures the kinetic energy of the blast and converts it into an electromagnetic field that is used to accelerate metal to much greater speeds than those achieved by simple blast. This can be done in a highly controlled fashion, so it can produce a more effective jet than a normal shaped charge, or multiple jets or fragments as desired.

Variants of MAHEM have been developed for wall-breaching, for destroying incoming artillery rounds, and for anti-tank use. But the most significant one is the EMEW (Electromagnetic Explosive Warhead) which will produce "Scalable Lethal and Nonlethal Effects."

Like the Star Trek Phaser, which can be set to stun or kill, by altering the way it detonates, EMEW can be a nonlethal "stun grenade" or it can scatter deadly shrapnel over a specific area. Additional settings would be used against heavy armor or to break through walls. EMEW is being developed by Enig Associates, with the aim of fitting to the new LMAMS devices. Their partner is Lockheed Martin, makers of the Terminator lethal drone. Meanwhile the Chinese are also working on the same MAHEM technology having apparently reverse-engineered the US concept.

Future small drone warheads will be equally lethal to vehicles, buildings, and people, and will present entirely new types of threat.

Flying Snipers

The concepts above are based on the general assumption that small drones would, like LMAMS and Switchblade, be sacrificing themselves in suicide attacks. Certainly they could be much cheaper than any existing munitions.

However, drones do not need to be expendable. Ravens have already been fitted-out as bombers. The US Navy has even developed a miniature GPS-guided bomb weighing about a pound, which a Raven-size drone could deliver from several hundred feet, from where it is invisible and inaudible.

In late 2015 Raytheon unveiled a new miniature laser-guided missile called Pike weighing just 1.7 pounds and with a range of over a mile. Drones armed with this type of mini-missile could be a game0changer on the battlefield, especially when teamed with other drones carrying laser designators. Pike is described as a fraction of the cost of the $70,000 Javelin missile; it is probably more expensive than the type of expendable drone discussed here, but as ever prices could fall sharply with mass production.

For the meantime, it makes sense for hand-launched tactical drones carrying high explosives to be confined to one-way missions. The risks involved with any drone returning with a live warhead seen as high, and this may be especially true when it has the slightly awkward, high-impact landing style of the Raven. However, new fuzing systems bring a new level of safety, and its certainly feasible that future drones may be reusable rather than being kamikazes. It is not hard to envisage an automated drone rearming station, situated just outside the combat area, with a stream of drone arriving to replenish their bombload. This would make small drones even more powerful.

Explosives are not the only option. A drone, even a small drone, might be armed with a gun, making it a miniature attack aircraft. A large version of this concept already exists. The Army's prototype Autonomous Rotorcraft Sniper System consists of a sniper rifle mounted in a stabilized turret fitted to a small helicopter. It will provide air support for urban operations, an eye in the sky that can also target enemies at will. It has "autonomous" in the name because an autopilot handles the tricky business of flying the helicopter, and the targeting system automatically compensates for windage and other factors. All the operator has to do is line up the target in the crosshairs on their handheld display. According to one of the developers it is as easy to use as an Xbox 360 controller.

Of course the ARSS is far too big for a small drone. It weighs over sixty pounds, and the hefty .338 Lapua Magnum rifle on its own weighs twice as much as a Raven. But a small drone does not need to have the ARSS's ability to place a bullet with pinpoint accuracy from a distance of one mile. If it can hit a target from a hundredth the distance - fifty feet away – it will still be an effective weapon.

In 2011 Kevin Jones and his team at the Postgraduate Naval School in Monterey experimented with a drone armed with a paintball gun, testing both quadrotor drones and a fixed-wing aircraft similar to a Raven. Their aim was to determine whether existing surveillance drones could be effectively armed with an add-on kit without losing the drone. As a typical mission, the armed drone might be able to distract a sniper to protect friendly troops.

They chose this armament not because it looked like a particularly useful weapon, but because it kept the paperwork to a minimum.

"The paintball round is actually more challenging [than a firearm]," says Jones, "in that the range and flight characteristics of the paint ball round are such that a bullet will almost certainly be more effective."

The paintball gun was to simulate a shotgun as an anti-sniper weapon. Accurate aiming from an unstabilized platform is difficult, but a shotgun's spread of shot could ensure that a sniper would be at least distracted and injured if not killed. From eighty feet it was difficult to hit the target with a paintball, but the near-misses were close enough that the target would have caught some of the blast from a shotgun. This type of attack would keep a sniper's head down and prevent him from targeting friendly forces even if it did not kill him.

Ideally, the operator would just have to click on a target and the system would maneuver into a firing position and aim the weapon, in the same way the Switchblade carries out the tricky terminal phase of the engagement automatically, but this was not practical, for the hardware available to Jones' team was fairly basic.

"It's always nice to offload operator workload to software," says Jones. "But without a gimbal and/or significant computer power on the aircraft for image processing, this would be quite challenging."

In any case, under current US policy a human will always be involved in the actual trigger-pull.

Jones carried out successful trials with both the fixed-wing and helicopter drones, but the work was not carried further, or at least not by his team.

While aiming was somewhat erratic, the experiment showed that a Raven armed with a suitable lightweight weapon would be able to engage a target. With better software and more processing power, it should be possible to hit the target with a rifle-type weapon.

Even a paintball gun could have its uses. Professional versions of the standard .68" caliber paintball weapon are used by police and others for "less lethal" rounds. These include pepper balls loaded with OC dust, which has the same effect as pepper spray, indelible marking rounds, and hard nylon rounds for breaking windows in hostage situations (they are usually followed by pepper balls or similar). A drone that can break windows would be able to access building interiors or open the way for others.

A paintball gun could also fire incendiary rounds, capsules of napalm or phosphorus mixture that burst into flames on impact. With a mix of window-breaker and incendiary balls, a single drone could start multiple fires inside many different buildings.

Jones is not the only one to have looked at armed drones. Firearms specialist McMillan Group International in Phoenix developed an eleven-ounce weapon firing standard 12-gauge shotgun cartridges or high-explosive airburst rounds. The mini-shotgun was originally designed for the Maveric hand-launched drone. It appears to have been intended for the same role as Jones' drone.

McMillan Group informed me that the original project was cancelled, but they declined to give details of later developments or the current status of the technology. Given that the Maveric is popular with US Special Forces, it is far from impossible that a flying shotgun is already in use. Unlike the Switchblade with its explosive warhead, this could return safely, firing off remaining ammunition before returning if necessary.

Snipe is a proposed armed version of the Silent Falcon solar-powered small drone, armed with an ultra-modern 5.7x28mm weapon. The small-caliber, high-velocity bullet is designed for highly compact pistols and submachine-guns. Developer John Brown believes the armed drone could hit man-sized targets at two hundred yards and could carry a magazine with fifty to a hundred rounds. Snipe has not yet flown but could be developed rapidly.

"If funding for the program was available, we would be about nine months away from having a prototype," says Brown.

There are other contenders. Australian outfit Metalstorm makes the Maul, claimed to be the lightest 12-gauge shotgun in the world at under two pounds. Maul is a five-shot design that fits under the barrel of a rifle and is used for door-breaking or non-lethal rounds. Maul has no moving parts and is fired electronically, making it especially suitable for installation on a drone as there is no risk of jamming as can occur with normal, mechanical weapons. Metalstorm has already tested their technology on small unmanned helicopters. While the Australian parent company went out of business in 2012, their technology is still available and Metalstorm USA was still trading as of 2015.

Lockheed Martin has looked at a multi-shot weapon as an expansion option for their Terminator LMAMS contender. The Terminator is a highly modular system, and future small drones might be able to carry either a shotgun or kamikaze explosive charge depending on the mission. In a cooperative swarm, the shotgun drones might be tasked with clearing away defenders.

There have also been homemade armed drones, "hacks" created by attaching a handgun to a commercial drone, with the results displayed on YouTube. They look crude but may still be deadly. New, small handguns like the Beretta Nano are chambered for 9mm, a standard military caliber, weigh about a pound including six rounds, and can have a laser sight built in for easy aiming. It is easy to see how this type of drone might be sent in to clear a building, reducing the need to risk soldiers' lives.

A drone armed with a firearm could carry out missions that currently take bombs. Again, it is a matter of accuracy being more important than brute force, especially when it comes to dealing with human targets. A five-hundred pound bomb may leave a target in a foxhole unscathed if it lands fifty feet away. A small drone that can perch on the edge of the foxhole, or hover above it with a shotgun, is far more deadly.

A persistent swarm of such drones would be effective at what the military call "area denial," keeping people out of a specific area. For example, it is not necessary to destroy a bridge to put it out of action. If the driver of every vehicle that approaches a bridge is attacked, the area will soon become impassable, clogged with vehicles that cannot be removed because of the drones. Equally important, people can see that the bridge is more dangerous than if it was strewn with landmines, and no driver would attempt to cross while the drone swarm is perched on and around it.

Existing drones can only release weapons when ordered to do so by their human controllers. It would take a large number of operators to handle the stream of targets that a swarm might shoot at one by one. However, as has already been mentioned, even mobile phone cameras now have people-detection software. They can distinguish a human in the scene and locate faces. It is only a slight step up to identify military uniforms, which are legally required to be recognizable and distinctive. A drone swarm could be unleashed and tasked with shooting every uniformed human in a given area.

Air to Air

Although small drones might be able to attack targets on the ground with some effect, air-to-air combat is another matter. No small drone can match the speed of a jet fighter, and in a dogfight they are going to be sitting ducks. Or more accurately, flying ducks. And flying ducks versus aircraft is an uneven competition.

Flocks of birds are a major hazard around airports, and both military and civil authorities go to great length to get them away from runways. The 2009 "miracle on the Hudson" occurred after an Airbus 320 ran into a flock of geese after taking off from La Guardia airport and lost power in both engines. Thanks to great skill and professionalism, the pilot ditched safely in the Hudson River and all passengers and crew were safely evacuated.

Drones are smaller than geese but may cause far more engine damage, especially if they contain metal parts. From the point of view of an aircraft, a drone swarm is a serious navigation hazard. A 2013 study by USAF Major John Mintz entitled "Asymmetric Air Warfare: A Paradigm Shift for US Air Superiority" makes this point emphatically:

"The speed at which an aircraft travels combined with the fragility of its jet engine intakes makes it an easy target for even low-tech, stationary swarms."

Aircraft approach paths to an airfield are fixed and impossible to defend. The same applies to aircraft carriers: a drone swarm in the air means that every take-off and landing brings a serious risk. And a one-pound explosive warhead in the jet intake would put any engine out of action; some modern surface-to-air missiles have warheads smaller than this. Even if the aircraft was not shot down, the loss of an engine would mean that a mission would be aborted at the very least, with the need for major repairs afterwards. A drone with a MEGOL-type system could maneuver into the most lethal position, targeting the cockpit, fuel tanks, or other vulnerable parts of an aircraft.

Ever since Vietnam, pilots have talked about the "Golden BB," a single bullet or small piece of shrapnel that could destroy a plane by striking a vulnerable spot. The difference with small drones a hit is not a matter of luck but aiming: not so much a Golden BB as a smart, precision-guided one.

Even away from airfields, swarms could present a danger to aircraft, a moving danger slower than the aircraft it stalks but able to shift and block access to particular areas, like a high-tech version of the old barrage balloon.

Persistent, swarming drones do not need to win in air-to-air combat. Being shot down in droves will not deter them. If a swarm loses ninety per cent of its aircraft and still arrives at an air base or aircraft carrier, it will be able to attack vulnerable planes on the ground, refueling or rearming. The manned aircraft soon suffer unacceptable losses; an exchange rate of one $100 million plane for a thousand $1,000 drones favors the drones by a hundred to one.

In practice, pilots would be as unlikely to take off into this sort of environment as foot soldiers would be to run through a minefield. The danger would be too great, and aircraft would be forced to remain in their hardened shelters or securely below decks on their carrier leaving dominion of the sky to small, buzzing machines.

A carrier or airbase under swarm attack is a besieged castle – and to those inside, it might feel very much like a prison from which there is no escape.

Non-Lethal but Dangerous

A suicide drone gets only one chance to attack, and a drone with a firearm has a limited supply of ammunition. But a weapon that can run off the drone's power supply can be used many times, especially if the drone recharge itself from power lines or solar cells.

Laser dazzlers or "ocular interrupters" are a good fit with drone capabilities.. They were deployed in Iraq and Afghanistan as non-lethal weapons, especially for dealing with drivers. Shining the brilliant green light on a car windscreen signaled to a driver approaching a checkpoint that they need to stop; and when you cannot see, you cannot drive. It does not cause flash blindness, but produces enough glare inside the eye so that it is impossible to see far enough ahead to drive safely. The exact effect depends on conditions, but typically a driver would only be able to progress at 20 mph at best. The dazzling laser also prevents the target from effectively aiming a weapon at the source.

The GLARE MOUT made by B E Meyers has been used extensively by US forces in Iraq and elsewhere. It weighs under ten ounces and is normally clipped on the underside of a rifle; effective range is four hundred meters at night and perhaps half that in daytime, even though the output is barely one-eighth of a watt. Aiming it is as simple as pointing a flashlight, and it would be simple enough to link it to a drone's camera. As with other weapons, it could be aimed wherever the camera detects a face.

Laser dazzlers are intended to be harmless to the eyes, but they can still be dangerous. The FAA notes that there are about eleven cases every day of pilots being targeted with laser pointers, and truck drivers and others have also been lasered, in some cases causing accidents. Drones with laser dazzlers could close a road by dazzling drivers, or spread havoc by flying down a freeway and dazzling at random.

Military personnel increasingly have special goggles providing laser protection, so fighter pilots ought to be safe – though this depends on having the right sort of goggles for the laser that's being used. Others may not be so lucky.

While the dazzling drone might seem to be little more than a nuisance on the battlefield, it makes an effective "pre-lethal" weapon. Anyone dazzled will not be able to defend themselves against more lethal attacks by other drones.

Another type of non-lethal drone has already been demonstrated. The Chaotic Unmanned Personal Intercept Drone (CUPID) seems a faintly ridiculous idea: a six-rotor helicopter drone armed with a Taser stun gun. It was created by Chaotic Moon Studios, a software company based in Austin, Texas, and demonstrated to the media in 2014. The idea is that the drone could autonomously patrol a given area or the inside of a building and detect intruders. It can call security personnel and, with their approval, go ahead and stun the intruder.

Modern Taser-type weapons require very little power. Early Tasers used several AA batteries, but the latest versions only need a couple of lithium batteries to give repeated five-second shocks. A drone equipped with this type of weapon can disable a human target for as long as necessary, for example to keep them out of action while the rest of the swarm completes an attack. Afterwards it can fly off and perform the same mission again and again.

CUPID was built by a software company with no particular expertise in this area – and without the knowledge or consent of

Taser Inc. They did it purely for publicity purposes, but the point was made. Now we know that anybody can build their own electroshocking drone.

Perhaps the most effective non-lethal weapons that small drones can carry will be jammers, which we will explore in more detail in the next chapter.

A Swarm of Stinging Bees

This chapter has only hinted at what the combined effects of a swarm acting cooperatively might be. Acting together, drones might bring down a bridge or a skyscraper, but they could do more than that. For example, while some drones start fires, others could block attempts to fight the fires. A fire truck is an easily identifiable vehicle and can be taken out by a small drone like Switchblade.

The firestorms of WWII were the result of blind mass bombing. A swarm incendiary attack would be more calculating and guided, picking optimal conditions and nurturing the growing blaze with additional strikes over a prolonged period.

The reader may speculate how effective a swarm would be against targets like oil refineries, chemical plants, or power stations – nuclear or otherwise.

The precision of a drone swarm means it can target particular aspects of infrastructure, such as communications or power supply. The effects of these are synergistic, amplifying each other: no phone signal and no power supply is worse than either singly. A city can become uninhabitable when the water, power, and sewage systems stop working.

As we have seen, the drone swarm has a persistence not matched by manned aircraft. Once the swarm arrives, it stays in place. Unlike air strikes by manned aircraft, the drone swarm does not go away again, with an all-clear announcing that the attacking aircraft have left the area. Perching or loitering drones will remain in the area for as long as there are targets to attack, even if they are temporarily hiding. You can hide in a bunker, but the drones will be waiting when you come out.

This air strike never ends. The psychological impact of this type of attack is impossible to assess, but it will represent a step-change in warfare.

Operation Ten-Go

Finally, anyone who feels that the killer swarm can be waved away as an improbable fiction ought to consider the disastrous outcome of Operation Ten-Go. As mentioned at the start of this chapter, before WWII, the battleship reigned supreme. There had been hints that battleships were vulnerable to air attacks, but navies were still based around capital ships. The Japanese took the idea to its ultimate extreme with the construction of two "super-battleships," Yamato and Musashi. At over seventy thousand tons these were the biggest battleships ever built, and the best-armed in the world, with nine eighteen-inch guns each.

As the war in the Pacific progressed, more ships were sunk by carrier-based aircraft than gunfire and the threat from their air became more obvious. Yamato's defenses were supplemented with over a hundred and sixty additional anti-aircraft guns.

When the Allies landed on Okinawa, the Japanese defense became increasingly desperate. The Imperial Air Force sent kamikazes. For their part the Imperial Navy sent Yamato, with eight escorting destroyers, on a one-way mission against the Americans called Operation Ten-Go ("Heaven One"). The battleship was to be beached off Okinawa so it could not be sunk and provide fire support for Japanese ground forces. It was a hopeless mission, and everyone knew it.

At 10 am on 7 April 1945, almost four hundred planes from eight American aircraft carriers converged on the Japanese force. The Japanese had no air cover, so the US planes were able to form up and attack at leisure. The first wave of planes concentrated on the screening destroyers, sinking them or putting them out of action.

The second and third waves attacked the battleship, hitting Yamato with at least eight torpedoes and fifteen bombs. The bomb damage was mainly superficial, but it destroyed the gun directors, the ballistic calculating devices used for aiming the anti-aircraft guns. The defenders were forced to aim guns manually, which made them far less effective. After multiple torpedo hits, Yamato's rudders were jammed and it started sinking.

Two hours in, the crew were finally ordered to abandon ship. It was too late; Yamato capsized and the ship was destroyed in a single giant explosion that left a mushroom cloud twenty thousand feet high. Less than three hundred out of the crew of three thousand survived.

From a military perspective, the most significant aspect of the engagement is not that the battleship was destroyed, but that it was destroyed so easily. Just ten US aircraft were lost in the assault, making the ratio of casualties on either side something like three hundred to one. A similar casualty ratio, in fact, to the brief Anglo-Zanzibar War.

Operation Ten-Go underlined the fact that air power now mattered more at sea than big guns. The battleship was destroyed by aircraft ten thousand times smaller than itself. Some would say that the point was already proven four years earlier at Pearl Harbor, when two US battleships were destroyed and five put out of action entirely by Japanese air power.

It will take a leap of imagination to grasp that large manned platforms are set to be eclipsed. They have not suddenly become useless, just as battleships continued to provide gunfire against land targets until the 1991 Gulf War. But the aircraft carrier and its squadrons of manned aircraft will become as outmoded as the old battleship in terms of combat effectiveness. A battle between a carrier group and a drone swarm would be as one-sided as Operation Ten-Go. It might not be as bloody, but it would certainly end with the naval aviators unable to carry out their mission. The world's most powerful weapon of force projection would be blunted to uselessness, like a battleship with its gun barrels plugged, an easy target for other weapons.

By contrast, drone swarms are likely to prove highly effective means of force projection. Whether delivered to the area by transport aircraft, surface ships, or submarines, or whether arriving under its own power, the drone swarm owns the area it occupies. The swarm operator can do exactly as much destruction as they like, to people, infrastructure, buildings, or military targets. They can shut down airports or block roads or railways at will. A modern Sultan of Zanzibar would find drones circling his palace and perched on the roof, ready to demolish the entire structure piece by piece until he surrendered.

One important point is that most of the technology described above – thermobarics and reactive materials – is not currently in the public domain. Unlike the various aspects of drone design discussed in Chapter 4, they could not easily be copied by non-governmental groups. In that sense, terrorists would be far more limited in the sort of payloads they could deliver by drone.

However, drones would still be a deadly weapon in terrorist hands. A five-hundred-pound car bomb outside a stadium does less damage than small drones with fragmentation warheads detonating twenty feet above the crowded stands. A suicide bomber cannot get into a government building, but drones can attack any room on any floor. Bulletproof glass may stop the first one or two, but not the rest.

Terrorists do not have the same restrictions as governments and might decide to arm their drones with chemical or biological weapons. Even a small quantity of Sarin or other nerve agent could cause mass deaths. Manufacturing chemical weapons is well within the capability of many groups. In 1995, terrorists released Sarin gas on the Tokyo subway, killing twelve people and harming about a thousand others.

This leads to the inevitable question: can we stop the drone swarm?

REFERENCES

1) During the Vietnam War the "circular error probable..."

Hallion, R.P. (n.d.). Precision guided munitions and the new era of warfare. *Air Power Studies Centre*. (http://bit.ly/1YLGRzl)

2) Thanh Hoa Bridge attack

Anderegg, CR. (2001). *Sierra hotel: Flying air force fighters in the decade after Vietnam*. Government Reprints Press

3) Schweinfurt ball bearing raids

World war II: Eighth Air Force Raid on Schweinfurt.(n.d.). (http://bit.ly/1RqhXTC)

4) B-1 Lancer strike targeting Saddam Hussein

Blair, D. (n.d.). Smart bombs aimed at Saddam killed families. *The Daily Telegraph*. (http://bit.ly/1Is2n84)

5) M150 Penetrating Augmented Munition (n.d.).

6) Tec Torch

Hambling, D. (n.d.). The ultrahot torch that slices through steel. *Popular Mechanics*. (http://bit.ly/1lzku1w)

7) Barnie and Bam-Bam reactive demolition munitions

Hambling, D. (2008, May). Reactive revolution: Meet the pulverisers. *WIRED*. (http://bit.ly/1XD0wnU)

8) ATK Reactive

Cvetnic, M. (n.d.). Reactive materials in mines and demolition systems. (http://bit.ly/1NsWM1H)

9) I-90 Innerbelt Bridge demolition

Grant, A. (2014, June). Thousands gather to watch early morning implosion of Inner Belt Bridge. (http://bit.ly/21zce2t)

10) SMAW-NE Thermobaric warhead

Hambling, D. (2005, November 14). Marines quiet about brutal new weapon. *Defensetech.* (http://bit.ly/1M7BtOr)

11) AGM-114N Metal Augmented Charge Hellfire

AGM-114N metal augmented charge (MAC) thermobaric hellfire. (http://bit.ly/1O67ojb)

12) XM0160 thermobaric grenade

Hambling, D. (2007). Thermobaric grenade brings down the house?. *WIRED.* (http://bit.ly/21zcqyH)

13) "A kind of flaming dew"

Guilan, R. (n.d.). The Tokyo fire raids, 1945. (http://bit.ly/1NsCNvC)

14) Project X-Ray bat bombs

Couffer, J. (2008). *Bat bomb: World war II's other secret weapon.* Austin: University of Texas Press

15) AT-4 Bazooka

M136 AT4, operation and function. (n.d.). (http://bit.ly/1QijUkd)

16) MEGOL

USAF SBIR description (now deleted). (n.d.). (http://bit.ly/1QijXMT)

17) AGM-114R

US Army briefing. (n.d.). (http://bit.ly/1QWWNgN)

18) Advanced Warhead technology (n.d.). (http://bit.ly/1OJLpBu)

19) MAHEM

Massey, K. (n.d.). Magneto hydrodynamic explosive munition (MAHEM). (http://bit.ly/1LRfmeq)

20) Autonomous Rotorcraft Sniper System

Hambling, D. (2009). Army tests flying robo sniper. *WIRED* (http://bit.ly/1NsFkWD)

21) Kevin Jones armed drone project

Hambling, D. (n.d.). The armed Quadrotors are coming. *Popular Mechanics*. (http://bit.ly/1QimzKK)

22) "The paintball round is actually more challenging" – Kevin Jones interview with author June 2012

23) Snipe armed Solar Falcon

Snipe (n.d.). (http://bit.ly/1NsYJeD)

24) "If funding for the program was available" – John Brown email to author

25) Metalstorm MAUL

Grieg, D. Metal storm completes first shoulder firing of MAUL shotgun. *Gizmag*. (http://bit.ly/1Is3XqG)

26) Homemade handgun drone

Lavars, N. (n.d.). Home-made handgun drone attracts FAA Investigation." *GizMag*. (http://bit.ly/1Q3ABRN)

27) Asymmetric Air Warfare: A Paradigm Shift for US Air Superiority

Mintz, J.P. (n.d.). Asymmetric air warfare: A paradigm shift for US Air superiority. (http://1.usa.gov/1Is47Oy)

28) Glare MOUT laser dazzler

Glare MOUT. (n.d.). (http://bit.ly/1HJ4Bjg)

29) CUPID

Halverson, N. Taser drone could stun criminals with 80K volts. (n.d.). (http://bit.ly/1O69zDx)

30) Operation Ten-Go

Hickman, K. (n.d.). World war II: Operation Ten-Go. (http://abt.cm/1OAGWmy)

CHAPTER NINE

FIGHTING THE SWARM

> *"They're Autons! Bullets can't stop them."*
>
> - Doctor Who, "Terror of the Autons" (scriptwriter Robert Holmes)

When William Meredith saw a small quadrotor hovering close to his property in Hillview, Kentucky, he was immediately suspicious. His two daughters were playing in the backyard, and he suspected the drone was spying on them, or looking for a house to burgle. Drones are already used by peeping toms and by burglars. Meredith got his shotgun. When the drone came over his property he let fly, easily bringing it down. The owner turned up soon afterwards, followed by police. They charged Meredith with criminal mischief and endangerment for firing into the air.

Cases like this make small drones look easy to kill. In discussions on the topic, people joke about using a flyswatter. The quadrotors flown by hobbyists are certainly vulnerable to the most basic of attacks, and YouTube videos show drones brought down by beer bottles, rocks, footballs, and other improvised projectiles. There is even a video of a chimpanzee knocking one down with a stick. As Meredith showed, guns make it even easier. Similar cases include hunters in Texas shooting down a drone flown by an animal rights group. .

While a hovering drone a few tens of feet away is an easy target, one circling a few hundred feet up is well above the range of any thrown missile and presents a challenging target for the shooter. A drone approaching at a hundred miles an hour – as Switchblade does – is virtually impossible to hit.

Hunters have difficulty hitting flying geese at more than about eighty yards, even with the spread of shot from a shotgun. Hitting one with a rifle is harder and putting a bullet through the Kevlar wing of a drone may only make it wobble. Unlike a goose or an airplane with "wet wings" containing fuel, a drone can only be seriously damaged by hitting a vital part.

A lethal drone like Switchblade will cover that last eighty yards to the target in around two seconds and its body presents a target four inches across. It can fly at low altitude, putting it below the horizon and making it difficult to see against a cluttered background. It can attack in complete darkness, and as it was seen in the section on swarming hunters, drones will come in from several directions at once. Some may even come from vertically above the target.

It may be possible to shoot down some of the attacking drones. But swarms are robust and casualties do not trouble them. As anyone who has ever faced a swarm of hornets will know, if even one gets to you, you have a problem. It is a much worse problem with lethal drones. Shooting down a single unarmed drone is easy. A swarm of lethal drones is another matter. If everyone who fires at the swarm is swiftly targeted and terminated, few people will be willing to risk their life by raising a gun.

Improvised weapons and small arms can only stop the unsophisticated drones, and then in limited numbers. A real swarm can only be stopped by organized air defenses.

In early 2015, the US Secret Service carried out flight tests in Washington, DC, to develop a new approach to dealing with potential attacks from small drones. The Combating Terrorism Technical Support Office – an organization set up by the Pentagon to deal with new types of terrorist threat – is also looking at exactly the same problem. Both are concerned that existing defenses will prove completely inadequate and are seeking new approaches.

Air Defense 101: Chasing Zeppelins

Air defense has a long history; as long ago as 1910 the US Navy was investigating bursting shells for use against enemy aircraft. This was a year before the first bomb was dropped from an aircraft in action.

Some doubted whether there really was an air threat. There was much debate in 1914 about whether air attacks would be legal. Since any attack on a city would harm innocent civilians, the general view was that it would not be acceptable under the laws of war.

The German High Command thought otherwise. They launched a series of raids on London with Zeppelin airships. Over five hundred feet long and with a crew of eighteen, the P-class Zeppelin looked more dangerous than it really was. It only carried three thousand pounds of bombs, and bombing from high altitude was a matter of pure luck at the best of times. Night-time bombing was effectively random; often the crew did not even know which city they were above but simply aimed at lights below. The immense craft depended on the wind being in the right direction, and missions were repeatedly thwarted by bad weather.

Fighter planes took a long time to climb to the altitude of the Zeppelin, and finding one in the dark proved to be surprisingly difficult. Ground-based defenses were needed. London's answer to the Zeppelin raids was a 75mm cannon mounted in the back of an armored truck. Being mobile, it could go out to meet the slow-flying attacker. Lt Commander Rawlinson wrote of having to negotiate London's crowded streets at high speed during one raid:

"I feel quite confident that no man who ever took that drive will ever forget any part of it, and particularly Oxford Street [London's busy shopping street], which presented an almost unbelievable spectacle.... I also observed several instances of people flattening themselves against shop windows, the public at that time being infinitely more fearful of a gun moving at such a terrific speed than they were of any German bombs."

When Rawlinson reached a suitable firing point, the aptly-name Artillery Ground, he stopped the truck and opened fire. The shell bursts failed to damage the Zeppelin, but may have helped drive it off. In any case, the event showed that even big, slow-moving targets were harder to hit than expected.

This was the start of air defense, a field marked by rapid evolution and improvisation with whatever weapons are at hand. In WW1 on the Western Front machine guns and a variety of light cannon were pressed into use, shooting at biplanes armed with machine guns and hand-dropped bombs. The cannon were mainly rapid-fire weapons firing a shell of one or two pounds. The results were not impressive, and when better aircraft were introduced, flying faster and higher, air defense became even more difficult. By the 1920s it was accepted that nothing could stop massed air attacks on cities and, in a famous phrase from the time, it was concluded that "the bomber will always get through."

By World War II there were a huge range of dedicated anti-aircraft weapons of all calibers, including the famous German 88mm Flak gun – Flak is short for Flugzeugabwehrkanone or "aircraft-defense cannon." As bombers became faster and flew higher, ever more powerful guns were needed, like the American "Stratosphere gun," firing a fifty-pound shell capable of hitting a plane at 57,000 feet – in theory.

In practice, a hit on a fast-moving target at long range was unlikely. Before the war, it was estimated that the guns would score one hit for every two hundred rounds. In reality it took closer to twenty thousand. A shell takes ten seconds or more to reach its target at high altitude, in which time a WWII bomber will have travelled about fifty times its own length. The slightest mis-estimation of range or speed means the shell has no chance of hitting. Anti-aircraft batteries fired a curtain of shells into the path of oncoming bomber formations rather than aiming individually. The mass of shell bursts did at least act as a deterrent. Later in the war proximity shells, which explode when they detect a plane nearby, greatly improved the chances that the shell would have at least some chance of doing damage.

Air defenses rarely shot down attacking aircraft. Shells did not hit planes, but sprayed them with high-velocity shrapnel fragments. The shrapnel generally caused minor damage or injured crew members, but this could force an aircraft to abort its mission and send it limping home. It took a lucky hit, or the cumulative damage from several near-misses, to down a plane.

A WWII survey found that 80 percent of injuries in the 8th Air Force were from shrapnel, which led to the development of the first flak jackets. These heavy garments were made of overlapping steel plates and saved many lives. The associated M5 groin armor was issued to seated personnel like pilots and was also very welcome.

On the tactical side, every tank and half-track had a machine gun for air defense, and Jeeps and trucks were fitted with machine guns in double or quadruple mounts to put up a wall of fire. Again, hitting a fast-moving target was difficult.

One approach was to make every fourth bullet from a machine gun a phosphorus tracer round that leaves a glowing trail. This showed the path of the bullets so the gunner could adjust his aim, directing the visible stream of bullets towards the target. Like the wall of shell bursts from larger guns, the stream of tracer was also a deterrent: it takes a steely nerve to deliberately fly into a hail of bullets.

For larger weapons, calculating sights and gun directors were essential. Again, shooting aircraft down was difficult, but a few bullets through the rudder or flaps, or damage to an engine or a fuel tank, would persuade an attacking pilot to break off his attack.

Kamikaze Menace

Air defenses were taxed to their limit when the Japanese resorted to suicide tactics in late 1944. Special Attack Units were formed of so-called kamikaze pilots who would crash their bomb-laden planes into American ships.

The kamikaze attacks were of two sorts. Sometimes single pilots would attack at spaced intervals, often aiming for ships that had been damaged by the previous attack. In some cases fire-fighting teams or damage repair crews were killed by the follow-up; in at least one case a kamikaze flew in through the hole in the side of a ship left by a previous kamikaze. On other occasions the Japanese carried out Kikusui ("floating chrysanthemums") assaults in which a large number came in at once from all directions.

Unlike other aircraft, the kamikazes were not deterred by slight damage. Machines guns and 20mm and 40mm cannon consistently failed to prevent a kamikaze from hitting his target. Only the big five-inch naval guns could destroy a plane with one hit.

"You have to blow them up, to damage them doesn't mean much," wrote Seaman First Class James Fahey in his memoirs of kamikaze attacks.

The military effectiveness of the kamikazes is still a matter of debate. One analyst calculates that, because they scored so many hits compared to the casualties suffered, kamikaze attacks cost the Japanese fewer planes per hit than other types of attack. Numbers are contentious, but around three thousand kamikaze attacks sunk or damaged some four hundred ships.

What is not in doubt is the psychological effectiveness of the attacks. Admiral Halsey, commander of the Third Fleet, described the kamikazes as "the only weapon I ever feared in war" after coming under attack at Okinawa.

"It looked like it was raining plane parts," wrote Fahey. "They were falling all over the ship. Quite a few of the men were hit by big pieces of Jap planes."

Newspaper reports were censored until April 1945, but the suicide attacks had a tremendous psychological impact on those who faced them. There were thousands of casualties of "combat fatigue," the name used for post-traumatic stress disorder, which was otherwise rare in the Navy. This psychological impact may be a foretaste of the effects of sustained attacks by drone swarms.

Admiral Halsey's solution to the kamikazes was an intensive program of air strikes on their airfields. Navy carrier air wings and Army Air Force B-29s destroyed large numbers of kamikazes on the ground, ending a threat that could not be stopped by anti-aircraft guns.

The End of Anti-aircraft Artillery

When jet engines were introduced in the late 40s, aircraft flew even faster and higher. Guns were clearly not going to be able to cut it for much longer.

The guided missile was the air-defense equivalent of the smart bomb. Instead of firing thousands of rounds and hoping for a lucky hit, a single projectile homed in on the target and guaranteed a shoot-down. Heat-seeking missiles were effective at close range, while bigger and heavier missiles with radar guidance took over at longer ranges.

In the 1960s, the US foot soldier had his personal air defense in the form of the Redeye missile. This was a portable heat-seeking missile that could take out a fast jet two miles away, an almost impossible feat even for a quadruple heavy-machine gun that had to be carried on a truck. The main problem with early versions of the Redeye was that it was purely a "revenge weapon" – it could only lock on to a jet's exhaust from behind, so you couldn't shoot down a plane until it had already flown over and bombed you.

In the same period, protection from heavy bombers was provided by the Nike Hercules. This missile stood forty feet high and flew at Mach 3 and had a range of eighty miles. While the Redeye carried two pounds of explosive, Nike Hercules was armed with a twenty-kiloton atomic warhead capable of bringing down a whole formation of bombers in one go.

The Army retained a few tactical anti-aircraft guns as a backup. In particular, missiles were of limited use at less than half a mile. Russia started to field squadrons of heavily armored Mi-24 Hind helicopter gunships in the 1970s. These were equipped with powerful guided anti-tank missiles, and threatened to decimate NATO armored divisions in the event of a war. NATO planners decided that a new gun-based air defense was needed for the modern age. It was to be called the M247 or Sergeant York after a famous US Army marksman.

The plan was to take the existing M48 Patton tank and fit it with a new turret armed with a pair of WWII-era 40mm guns. Manual aiming was not enough; it would be guided by the radar from an F-16 aircraft with a new computerized fire-control system. On paper, the Sergeant York looked like a sound proposition.

The result was a billion-dollar fiasco. The Patton tank chassis were worn out, giving up after three hundred miles of road tests instead of the four thousand planned. The 40mm guns had been stored badly and were in poor shape. The biggest defect was the radar; designed for air-to-air combat in the open sky, it could not deal with all the clutter at ground level. It was easily confused by things like waving trees, which it mistook for helicopters. In one test, a drone fighter had to fly over seventeen times before it was hit.

In another test the Sergeant York turned away from a drone target and aimed its guns at a latrine fan, which it mistook for a set of rotor blades. Test targets were augmented with radar reflectors to make them stand out better and ensure the radar could lock on. One commentator called this "demonstrating the abilities of a bloodhound by having it find a man standing alone in the middle of an empty parking lot, covered with steaks."

After six years and $1.8 billion, Sergeant York was cancelled.

These days US Army's only anti-aircraft gun is in the AN/TWQ-1 Avenger. This is a Hummer with an air defense turret carrying eight Stinger missiles and a solitary .50 caliber machine gun. The bulk of air defense is squarely on the missiles. But what missiles!

The Ultimate Missile

While anti-aircraft guns have shuffled into virtual extinction, missiles have flourished. They have changed little externally. The modern Stinger looks a lot like the 1960s Redeye, and the Patriot missile looks like a smaller version of the old Nike. Rather than being bigger and more powerful, they are smarter and more agile. As with bombs, intelligence trumps brute force.

Modern missiles can spot targets faster and shrug off the clutter that confused Sergeant York. They are highly resistant to jamming and deception. They are harder to avoid in the dance of death known as the "terminal engagement phase," when planes maneuver wildly in a desperate attempt to get away as the missile closes in.

Air defense has become a duel between radar operators and "defense suppression" aircraft equipped with electronic warfare pods, decoys, and missiles that home in on radar emissions. The attackers attempt to blind, confuse, or evade the defenses and get close enough to launch their missiles. A radar signal is like a searchlight on a dark night, advertising its position over a wide area. Radar operators respond by only turning their radar on at intervals, and by moving position when possible. It is a duel whose outcome is largely determined by who has the best technology.

The current refinement of the Patriot missile is state of the art. This is several generations on from the missile that was hailed (inaccurately) as the Scud-buster of the 1991 Gulf War. The fifteen hundred pound missile travels at almost a mile per second and can destroy an aircraft anywhere from treetop height to eighty thousand feet, at a range of a hundred miles away. Costing somewhere over a million dollars per shot, the Patriot is an effective weapon against a whole range of targets. A battery of Patriots can defend against attack helicopters like the Hind, strike aircraft, heavy bombers, and is now effective against Scuds and other ballistic missiles.

The recent focus has been on tweaking Patriot for missile defense because shooting down aircraft simply is not an issue. US air superiority in recent conflicts means that nobody has been in a position to bomb US forces. According to the USAF's 2014 Posture Statement:

"Since April of 1953, roughly seven million American service members have deployed to combat and contingency operations all over the world. Thousands of them have died as they fought. Not a single one was killed by an enemy aircraft. We intend to keep it that way."

In the near future the situation may be different.

Missiles versus Drones

Swatting down large slow-moving, non-stealthy drones like the Predator or Reaper is child's play to a Patriot. As we have seen, a major criticism of large drones is that they are no use in defended airspace. A drone swarm is another matter. The sharp end of a Patriot missile battery comprises four launch vehicles, each with four missiles ready to fire. In principle, a Patriot battalion can take on sixteen aircraft at a time. While two or more missiles may sometimes be launched on different trajectories at a difficult target, the battery might take out sixteen Reapers in a matter of seconds.

Whether Patriot could even hit small drones is another question entirely. The military are not willing to discuss how small a target a Patriot could lock on to, and whether a hand-launched drone would even be visible to it. It is hard to image a three-quarter ton missile engaging a four-pound drone. And even if every missile worked perfectly, the seventeenth drone would get through -- along with all those following.

Patriot missile batteries rely on radar, which is vulnerable; one hit could put the whole battery out of action. The drones might target the launch vehicles and personnel. Systems like the Patriot are not armored against attack, and the M983 trucks that transport the Patriot are as vulnerable as any other truck. Missiles are explosive targets full of flammable rocket fuel.

You cannot overwhelm a Patriot battery with a manned air strike, not unless you have a whole squadron of pilots willing to sacrifice their lives. While such units may exist in Iran's Revolutionary Guard, they do not generally feature in modern air warfare.

Drones, on the other hand, are immune to casualties. Using Patriots against small drones would be like trying to stop a swarm of angry hornets with a sniper rifle. A swarm of hundreds or thousands of drones would be simply unstoppable.

In World War II, the response to the tactical air threat was to issue a large number of machines guns, both on land and at sea. But machine guns are of limited use against drones – even if you see them in time.

Nor can the problem be solved by issuing Stingers to every soldier; at over $38,000 a shot, they are too expensive to be bought in such volumes. Worse, missiles like the Stinger are heat-seekers that depend on the target having a hot engine. A small drone with an electric motor is invisible.

As we saw in the last chapter, air power is equally helpless against a swarm. The USAF's F-22A Raptor is arguably the best fighter in the world, but its six radar-guided AMRAAM missiles and two infrared Sidewinders will not dent a swarm, even if they were able to lock on. The Raptor's 20mm cannon makes little difference. The rotary cannon has a high rate of fire to ensure a good chance of a hit, and the entire magazine is expended by six one-second bursts.

Whether a fighter could hit the tiny Raven is open to question. Because of the limits of stall speed, the Raptor pilot would approach at a relative speed of over a hundred miles an hour, giving little time to line up a shot.

It is doubtful whether a pilot would risk catastrophe for the sake of slowly chipping away at a drone swarm. The dangers to aircraft outlined in the previous chapter indicate why no fighter pilot would want to get near them.

Against most opponents, air supremacy means destroying enemy air fields so their aircraft cannot take off or land. This was the answer to the kamikaze threat. Hangars, even hardened ones, can be targeted. The enemy air force can be destroyed on the ground unless it takes drastic measures, like the hundred or so Iraqi Air Force jets that fled to Iran in 1991.

Small drones do not need a runway, air base, or hangars. If they can be attacked while they are being stored or transported, it might be possible to destroy large numbers in one go. But once they have been launched they are all but immune to conventional air power.

The Secret World of Black Darts

The Pentagon is fully aware of the threat from small drones, but its plans for dealing with them have been kept secret. We know plenty about anti-missile, anti-aircraft, anti-submarine, and other weapons, but little about what defenses there are against drones.

Black Dart is an annual exercise held since 2010 to evaluate anti-drone tactics. It is a joint exercise with Army, Navy, and Air Force personnel participating. Little is revealed about the equipment used, the targets they are tested against, or how good (or bad) the results are.

Black Dart concentrates on Class 1 and 2 drones, the smallest types which are likely to be problematic for existing defenses. The public face of Black Dart in 2014 included a display of radio controlled model aircraft to show possible threats, from a fifty-pound radio controlled jet – a scale model of an F-84 fighter – to quadrotors weighing a few ounces.

Some fourteen hundred participants attended Black Dart 2014, with eighty-five "systems," including weapons, sensors, and electronic devices such as jammers. Eleven types of manned aircraft were involved in exercises, along with a US Navy cruiser. For targets they had thirteen types of drone, from the Global Hawk with a wingspan over a hundred feet to the fourteen-inch Bandito, which can be launched from larger drones. The big drones were there to test target detection and tracking, but a number of the smaller types were shot down, falling victim to the cruiser's rapid-fire Phalanx cannon and five-inch gun. Details, however, are scant.

Some degree of secrecy is understandable. You do not want potential opponent to know what tactics are most effective either for small drones or those attempting to counter them. If Black Dart finds that dozens of drones attacking from multiple directions are all but unstoppable, it would not be good to advertise this to enemies.

However, the secrecy also leaves taxpayers in the dark about drone defenses. The embarrassment over the success of Firebees against US air defenses meant that few people were aware just how effective unmanned aircraft were. And perhaps the biggest lesson the military learned from the Sergeant York debacle was not to publicize its failures too loudly.

Counter-UAS

Having used the Raven and Switchblade successfully themselves, the US Army knows it is only a matter of time before they face an opposing force with similar capabilities. And they recognize the gap in their defenses.

The US Army's Armament Research, Development and Engineering Center (ARDEC) has a Counter-Unmanned Aerial Systems project to fill the gap. It is what is known as a "system of systems," involving a variety of different sensors, a central control system, and various weapons. The idea is that all of these can be plug and play, so a new acoustic sensor or a new missile can be added to the mix when it is available. Whatever else, it will need to be cheap so it can be issued in large numbers.

"ARDEC's focus is on the development of an affordable close-in counter-UAS system that is capable to defeat smaller sized UASs," says ARDEC's Project Officer Hannibal People.

Money is tight, and the C-UAS developers know they would not get funding for a whole new suite of weapons to fight off small drones: "We look to utilize existing weapon systems that can address this threat while maintaining capability against their conventional target sets."

Three different anti-drone capabilities have been demonstrated. In one of them, sensors detected a drone and passed targeting information to a remote-controlled turret, automatically aiming it at the threat.

Many small vehicles now have remote-controlled turrets, known as CROWS or Crew-Operated Remote Weapon Stations. They were introduced in Afghanistan and Iraq when the previous approach of having a machine-gunner on top of a vehicle proved too dangerous. Sitting half-out of a vehicle meant that the gunner was not protected by the vehicle's armor, and many were injured by roadside bombs. The billion-dollar CROWS effort provided Hummers and other vehicles with powered machine-gun turrets with video cameras and thermal imaging, operated from safely inside the vehicle. These turrets do not even need a soldier to operate them: because it's controlled electronically, the gun can be aimed as easily by a computer as by a human. At present though, a decision to fire must always come from an operator.

This raises the spectre of Sergeant York. Taking existing vehicles and weapons and making them work with new sensors and new control systems sounds simple in theory but can be difficult in practice. In addition, having automated guns that choose their own targets, a virtual necessity in the heat of a drone swarm attack from all directions and all angles, may cause problems. Considerations about friendly fire loom large. If the price of an anti-drone defense is soldiers lost to fratricidal fire, then it is not going to be an effective solution. Soldiers will not turn the weapon on if it may shoot one of their own.

A second version employed a new "gun-launched munition," a term used to describe a guided projectile fired from a cannon. This is initially being tested with a 50mm round, a caliber not widely used by US forces, but if a smaller version is developed it might give anti-drone capability to vehicles like the M2 Bradley, which mounts a 25mm cannon. This type of round is more likely to be of use against larger drones, but would be lethal enough against Raven-sized targets.

Another strand of the counter-UAS effort developed by the US Navy is based on the Navy's Spike missile. Described as being the size of a baguette, this is intended to be the smallest and cheapest guided missile in the world, made largely from off-the-shelf components. The design philosophy is remarkably similar to the MITRE team's approach to their Razor drone, rejecting the expensive, "exquisite" approach of customized electronics to make something that can be cheap and numerous. For example, it uses a video camera rather than an expensive laser seeker for laser guidance.

Spike weighs five pounds and has a one-pound warhead, making it the same size as the threat. (The newer Pike missile mentioned previously is smaller but likely to be more expensive) Directed by the same sensors and control system as the machine-gun based defense, Spike would provide a longer-range and deadlier alternative with the effectiveness of a smart weapon. At just $5,000 a shot, it is a fraction of the cost of the Stinger and other modern air defense missiles, and cheaper than current military drones. Given sufficient numbers, it looks like the most feasible approach to dealing with swarms.

Many questions about the C-UAS cannot be answered for security reasons, indicating the Army is taking the threat seriously. The project might take some lessons from the recent C-RAM program, which aimed to protect troops from rockets, mortars, and artillery. But while the C-RAM threats are easily visible from a distance and travel in predictable trajectories, enemy drones might not move in a straight line. Rockets, mortars and artillery shells usually only arrive a few at a time; the great problem with the swarm is that there may be hundreds or thousands of targets all appearing simultaneously.

The earlier C-RAM only aimed to protect installations. C-UAS will have to protect soldiers out in the field, which is likely to prove more challenging. It is not quite as difficult as chasing Zeppelins with a cannon mounted on the back of a truck, but it shares the assumption that the tools we have now will just have to do the job. But there are alternatives.

Directed Energy

It is not easy to hit a drone with cannon or machine guns, even with radar-guided, computer-controlled aiming. But hitting one with a laser is much easier.

Laser weapons have always had a certain appeal to the military mind, since before they even existed. The heat ray in HG Wells' 1901 *The War of The Worlds* looked like the future of warfare. A weapon that struck at the speed of light and could be aimed as easily as a searchlight was inherently more accurate than any gun. With lasers, there is no need to calculate the lead even with a fast-moving target. A jet moving at Mach 2 travels about an inch in the time it takes for a laser beam to reach it from ten miles away. From the very first days of lasers, the Pentagon has been trying to turn them into weapons.

Although lasers are all around us, from DVD players to laser printers and broadband connections, laser weapons failed to materialize. The problem was the size of the laser needed to produce a powerful enough beam. One researcher joked that the best way to kill someone with a high-power laser was to drop it on them.

But while destroying a tank or an aircraft is out of their league, lasers have long been able to destroy drones. In November 1973 the USAF gave one of the earliest demonstrations of laser weaponry, shooting down the usual hapless drone over New Mexico with a carbon-dioxide gas dynamic laser. Since then the demonstrations have continued at regular intervals, with Firebee drones being frequent targets. The lasers have always been too large and too fragile to make practical weapons, or have had other drawbacks.

That is changing. Mobile laser weapons are currently in the range of tens of kilowatts. Unlike earlier lasers powered by chemical reactions, they are electric, so can keep firing for as long as they have power, giving them an effectively unlimited magazine. They are not powerful enough to burn through armor but are capable of destroying missiles or small drones.

The great thing about lasers versus small drones is that the cost-per-shot is so low. Shooting down a $1,000 drone with a $5,000 missile is not a winning strategy. A $1 burst of precisely-guided laser energy makes much more sense. Also, the laser does not have a limited ammunition supply, but can keep firing as long as the generator has fuel. In principle, it can keep firing for as long as the drones keep coming, though lasers still tend to overheat after a while.

There are several such systems under development. The US Navy's Ground-Based Air Defense Directed Energy On-the-Move (GBAD) is typical. This aims specifically at providing "an affordable alternative to traditional firepower to keep enemy unmanned aerial vehicles from tracking and targeting Marines on the ground." (The Navy develop systems for the Marine Corps).

The GBAD system will be carried by Hummer-sized tactical vehicles as a defense against enemy drones. As of 2014 it has a power of ten kilowatts, but a full thirty-kilowatt system should be available by 2016. A short blast from this should be able to "disable, damage or defeat" a drone. Even if it does not destroy the drone outright or cause it to crash, the laser will burn out optics and damage sensitive control surfaces or other components.

Meanwhile the US Army's own truck-mounted High Energy Laser Mobile Demonstrator (HEL MD) is continuing along a gradual development path. This system is starting at ten kilowatts, but will be built up to higher powers capable of tacking a wider range of targets.

A larger system is already in use. The US Navy's prototype Laser Weapon System (LaWS) has been installed on the USS Ponce and will be working alongside the existing Phalanx Close-In Weapon. LaWS is effectively six industrial welding lasers with beams converging on the target and will be used against drones, cruise missiles, and small boats.

The Chinese are thinking along the same lines. In November 2014 Xinhua News Agency carried a story about a new Chinese laser system specifically designed to tackle small drones. It has a range of a mile and can bring down a small drone within five seconds. Officials acknowledged the threat from terrorists using drones and stated that the new laser would "play a key role in ensuring security during major events in urban areas."

The laser was built with the help of the China Academy of Engineering Physics and is effective against drones moving at up to 110 mph.

Lasers ought to be far more effective than guns or missiles for shooting down small drones. However, the engagement is likely to be a short-range one. Although the laser may have a range of a mile or more, as soon as it is spotted or starts firing, the drone swarm is likely to drop low and hug the ground for cover, limiting the laser's effective range to a few hundred yards at best.

It then becomes a question of just how rapidly the laser system can detect, track, target, and destroy the incoming drones. This will determine just how many members of a swarm the laser can down before it is overrun. If it starts at a few hundred meters, it will be less than ten seconds before the drones are at point-blank range. Given those figures, the Chinese laser might get two or three drones before it starts getting hit.

If effective, laser systems may provide a bubble of defense where they are deployed, protecting ships, forward operating bases, and convoys. If they can be produced cheaply in large numbers, then they might be able to provide an effective umbrella...but even so, the umbrella could still be overwhelmed by a big enough swarm.

This scenario assumes that drone developers simply ignore the threat from lasers; in practice they will introduce countermeasures. High-energy lasers operate on a single wavelength, so anything that reflects or absorbs that particular wavelength may reduce its effectiveness. The laser defense may be defeated by something as simple a mirrored nosecone, although this is not nearly as easy as it sounds. The reflective surface has to be tailored to the type of laser it is facing. What works against a laser working in visible wavelengths will not be effective against an infrared or ultraviolet laser.

Laser protection does not need to be absolute. Protection that means that each drone takes several seconds rather than one second to destroy will guarantee success for the swarm.

Laser countermeasures are already on the way. In 2009, the Navy issued requests for a laser-protection system for small drones that occupies less than fifteen cubic inches and could be produced at a cost of less than a thousand dollars per set. Adsys Controls of Irvine, California, won a contract for a protective device that works by rapidly detecting the incoming laser and deploying "novel countermeasures to disrupt the DEW [Directed Energy Weapon] tracking mechanisms." This might be something like an aerosol producing a reflective cloud to confuse the laser tracking system. Interestingly, the Navy wanted the countermeasures to be fitted as an add-on to existing drones. Development continued into a second phase through 2012 and the system may now be operational.

Small drones evolve fast. It takes many years to develop, manufacture, and get a new mobile laser system into service, but drones can be modified in months or less. The entire fleet does not need to be upgraded; a swarm only needs enough laser-proof members to deal with any laser defenses it encounters.

GBAD is costing somewhere over $22m to develop, and the individual units are likely to be expensive if and when they are fielded. When it comes to the numbers game, the swarm of cheap drones which can soak up casualties wins every time. And if lasers do ever become small and cheap, then laser-armed drones will be an early application.

Electromagnetic Pulse Weapons

While lasers may not be able to destroy drones fast enough, another sort of futuristic directed energy weapon might be able to exploit a drones' Achilles Heel: their electronics. An electromagnetic pulse (EMP) or high-powered microwave beam produces strong electric currents in electronic components. If you put metal foil in a microwave oven, you will see some dramatic sparking – and destroy the oven. The same sparking is what makes an EMP so effective against electronics. At the extreme end it can literally burn them out and render them useless; at lower levels it will cause temporary failures and glitches so the device needs to be rebooted.

One problem is that EMPs destroy all friendly electronics within range as well. The movie Matrix Revolutions, part of a series not noted for scientific accuracy, did at least get this one right: the protagonists used an EMP weapon that wiped out the first wave of an attacking drone swarm but left them even more vulnerable for the next one.

There is little information about EMPs and high-powered microwaves in the public domain. Although there have been rumors of tests in Iraq and elsewhere, no EMP weapons have ever officially been used in action.

EMP devices come in two varieties: One sort resembles radar, with a narrow, powerful beam of radio waves. The other is a bomb that converts explosive energy into a radio pulse emitted in all directions.

In October 2012, the Air Force Research laboratory announced the test of its Counter-electronics High-powered Microwave Advanced Missile Project (CHAMP). Developed by Boeing's Phantom Works, this is mounted on a cruise missile and concentrates its energy in a narrow beam to degrade or destroy electronics. A video of the event, called "Lights Out," showed computer screens going blank as CHAMP was activated. CHAMP is capable of carrying out multiple attacks on different targets, taking "tens of seconds" to recharge itself between zaps.

CHAMP looks like an effective way of knocking out communications centers, command and control, or indeed most of the functions of modern civilization. One pulse and the enemy is sent back in time to the nineteenth century with no computers, cars, televisions, telephones, or any of the other gadgets we tend to take for granted.

CHAMP is not quite ready for action yet. A follow-up program, SUPERCHAMP, is needed to extend its capability to "operationally relevant ranges."

There are also much smaller weapons. The US Army is working on an EMP device the size of a hand grenade.

Explosive devices that produce EMP may cause damage for tens or hundreds of meters. It is impossible to say exactly what it would do as the effects on a given target are inherently unpredictable. It is not just that different components fail at different power levels; the effect of the pulse depends on the angle at which it strikes the target. In the days before cable and satellite, television owners found themselves waving their television aerial around to get a signal, because the quality of the picture depended on pointing it exactly at where the signal was coming from. The situation with EMP is similar, and the pulse may cause severe damage or none at all depending on which way the target is facing.

The effect also depends on the arrangement of components in the target. Modern cars, for example, rely on electronic systems and can be disabled by a suitably powerful microwave pulse. The problem is that the strength of pulse required varies wildly between different cars. This has hindered the development of microwave-based weapons for the police to stop fleeing vehicles. A device which was powerful enough to knock out the most resistant vehicles at close range would disable other vehicles several blocks away, causing an unacceptable level of collateral damage.

The military are well aware of electronic friendly fire. The USAF's current policy is that this type of weapon should only be mounted on unmanned aircraft or missiles because there is too much threat that a manned aircraft could inadvertently damage its own systems.

Some types of radar may be powerful enough to qualify as microwave weapons. The latest upgrade to the Active Electronic Steered Array radar on the F-22 and F-15 fighters, known as APG-63(V)2, is said to be powerful enough to damage the guidance electronics on incoming cruise missiles at close range. This might be a better way for a fighter to tackle drones, as it has an unlimited supply of ammunition and can zap them one after the other. Its effectiveness will depend on what sort of range it has and how long it takes to have an effect. The main problems are likely to be the same as for using cannon: getting close enough for long enough when there is a speed difference of at least a hundred miles an hour, and avoiding running into other drones while doing it.

It is possible to shield electronics against EMP by placing them inside a conducting "Faraday cage" and ensuring that any external receivers such as antenna are protected. This is easier with a small device, such as the control system for a small drone, than a large one. In addition, the dispersed nature of the swarm means the geometry of the attack will only be favorable for some of them, and much of the swarm is likely to survive a single pulse.

EMP and high-powered microwaves may be useful for taking out small drones, but again this looks like a weapon which, when fielded, will work more in favor of the swarm. Radar inevitably gets smaller, and a swarm of drones with radar powerful enough to damage electronics could sweep across a city and selectively knock out sections of the infrastructure.

A drone can carry a hand-grenade-sized EMP weapon, allowing it to do a disproportionate amount of damage to a military headquarters, or knock out a tank or aircraft more easily than explosives. Against infantry it would leave a squad of soldiers without their communications, navigation, or night-vision equipment. An EMP device large enough to stop a whole swarm of drones would do massive damage to the infrastructure – and if the drones are well shielded, the defender would be the loser.

Jamming the Drones

Being without a pilot gives drones certain natural advantages. They can be made small, cheap and expendable. But having no pilot means drones are basically dumb. They may be able to carry out some missions on their own, such as taking pictures of a designated area, but the current generation have limited autonomy. For any sort of attack, the Pentagon insists on having a man in the loop.

This makes drones reliant on radio communications. The Raven is limited to more or less line of sight, though this can be extended by the use of relays. The Predator and Reaper can span the globe but need to maintain a satellite link to their controller. Break the link between the operator and the drone and the drone becomes useless, returning to base or roaming around trying to get a signal.

Ever since radios have been used in warfare, there have been efforts to interfere with them. Britain's Royal Navy experimented with broadcasting signals to interfere with enemy communication as far back as 1902, just five years after the first radios were installed on ships. By WWI, the use of radio was more sophisticated, and so were countermeasures. The German Zeppelins used fixes from radio stations to navigate; so on the night of 19 October 1917, the French switched radio broadcasts from the Eiffel Tower to another station to send the airships off course. When the Germans started using their own transmitters for navigation, the Allies drowned these out with louder transmissions on the same wavelength, possibly the first attempt to deliberately jam radio reception.

Electronic warfare has continued ever since. It can include radio direction-finding to locate an enemy, signals intercept to listen in to their communications, signals analysis to detect where major units are by the volume of traffic, "spoofing" with false messages, and of course jamming. As soon as radio-controlled weapons first appeared, the jammers were ready for them.

The first radio-guided weapon to see action was the German FS-1400 or Fritz-X developed in WWII. This was a three-thousand-pound glide bomb designed to attack heavily armored battleships and cruisers. It was a simple bomb fitted with small fins worked by radio control. A flare on the tail of the bomb allowed the bomb aimer to follow its progress and adjust its course with simple up-down, left-right corrections.

Decades before laser-guided bombs, the Fritz-X was a potentially revolutionary weapon that would allow heavy bombers to score direct hits on moving ships from high altitude. The weapon proved itself in the Sicily campaign in 1943 when the Italians went over the Allies: it sank the Italian flagship Roma and damaged the British battleship Warspite and several other Allied warships. This success was to be short lived. The Allies captured Fritz-X missiles and control equipment, and had developed effective countermeasures within a matter of months.

The British Type 650 Transmitter, and jammers from the US Naval Research Laboratory and Harvard's Radio Research Laboratory, nullified the Fritz-X. The jammer steered the bomb as far over in one direction as possible, overriding the operator's commands. From being a guaranteed hit, it became a guaranteed miss. The Germans dropped the idea of radio guidance, and the next version of the Fritz-X was guided via a wire spooling out of the back of the missile.

There were a few other radio-guided missiles, including some early Russian anti-tank guided missiles, but again these were too easy to jam. Until the advent of laser, anti-tank missiles were wire-guided and anti-aircraft missiles were heat-seeking or radar-guided.

At its simplest, jamming may simply mean broadcasting noise in the frequency band that the receiver is operating in. This sort of brute force jamming is rare in military circles, where communications tend to hop from one frequency to another at rapid intervals, and it is difficult to jam the entire spectrum with enough power. Smart jammers detect and analyze an opponent's communications and can selectively jam only in the ranges where needed. Jammers are also likely to be directional, rather than blasting out noise in all directions.

In the Iraq and Afghan conflicts, tactical jamming took on a new urgency. Insurgents had started triggering bombs using cheap cell phones. Special countermeasures were fielded to block the signals; the Pentagon spent some $17 billion on electronic countermeasures with some fifty thousand jamming units being issued. These included portable Warlock Green units for foot soldiers, which are credited with saving many lives.

The same technology could be used against swarming drones, especially ones relying on off-the-shelf cell phone technology. It is notable that the 2014 Black Dart exercise included an EA-18 Growler, the most modern electronic warfare aircraft in the Air Force's inventory, equipped with a range of powerful jammers. When the radio signal to a drone is jammed, it is usually programmed to return to the last point where it could communicate or simply return to base.

Some systems are already being designed with this in mind. The Anti-UAV Defense System unveiled in early 2015 comprises various sensors linked to a radio frequency disruption system produced by British company Enterprise Control Systems. According to the makers it can "disrupt and bring down any malicious drone in a phased and controlled manner" from more than a mile away. It was quickly joined by other anti-drone jamming systems: Icarus from Lockheed Martin, Selex's Falcon Shield and a device from Battelle called DroneDefender which looks like a ray gun from a science-fiction B-movie. With all of these devices, detecting and locating the incoming threat is the limiting factor, which is why many are linked to radar and other sensors.

This approach will certainly take out domestic drones which are not designed resist jamming. Military drones have a degree of resistance; the Reaper has a satellite dish pointed upwards, away from ground-based interference. As we have seen, swarms can be resistant to jamming, and the structure of a mesh network means that each drone only needs to communicate as far as the next. If one of the swarm acts as a gateway with communications to a satellite or other external link, the whole swarm can be controlled remotely.

More sophisticated electronic attacks may hack into drone control systems. This is a vulnerability of commercial drones that do not use encrypted links and work on known frequencies. Even military drones may be hacked; the Iranians claim that the RQ-170 Sentinel drone that came down over their territory was brought down deliberately, although a systems failure is just as likely. The Combating Terrorism Technical Support Office and others are looking at hacking to safely deal with consumer drones repurposed by terrorists.

More advanced drones will be harder to hack. Raytheon has already demonstrated a system called Electronic Armor intended to make small drones hack-proof all the way down to quadrotor size. It is an arms race between the jammers and the communicators. So far the advantage has been with the communicators.

However, future drones may not rely on radio at all. There is a growing interest in free-space optical communications, sometimes described as "like broadband over optical fiber, but without the fiber." The laser signal is beamed through the air to the receiver. This method only works over line of sight (obviously enough) and, because of atmospheric effects, the range tends to be limited to a mile or so. It can carry as much data as a broadband fiber optic cable and would be ideal for a swarm of drones forming a mesh network. Because it does not rely on radio waves, optical communication cannot be jammed or hacked into.

In a more low-tech version of optical communication, researchers at the Postgraduate Naval Center in Monterey have looked at communicating via QR codes as a form of "digital semaphore." These are two-dimensional bar codes, familiar from adverts, where you are urged to scan the code with your smartphone to go to a web page. In their experiments, a drone flashed up QR codes on a display panel like a smartphone screen; these were read by another drone. The team found that QR codes could be read from over five hundred feet away. A swarm could pass messages between members to coordinate its actions.

The bandwidth for digital semaphore is limited, but swarms like Lockheed Martin's SPAN in the previous chapter are designed to work with minimal bandwidth. It does not take kilobytes to convey the message "converge at XXX coordinates at time XXX to attack target XXX" through the entire swarm. In an earlier age, fleets of ships were marshalled effectively by flag signals and an earlier version of semaphore.

As we have seen, a few simple swarming behaviors mean that flocks of birds, foraging bees, and hunting wolves hardly need to communicate with each other. Simply observing the location of the nearest members of the pack, flock, or swarm is enough for them to coordinate their actions effectively. Trying to jam a swarm of drones may be as ineffective as trying to jam a swarm of locusts.

It may be argued that jamming communications will stop a drone swarm because they still need to communicate with their human operator. This is true for real-time reconnaissance, the "Predator crack" of a live video feed, but a drone can always fly back out of jamming and send back data later. And for attacks, drones do not need human involvement. While the US military currently insists on keeping a man in the loop for drone attacks, others may not. They may view unleashing the swarm as more like firing a missile: point it at the target and let it destroy what it finds there.

There are some fine legal arguments about exactly how much control is needed. An artillery shell is out of control as soon as it leaves the barrel, as is a heat-seeking missile when it leaves the rails. Some missiles, designed to target air defense radar, already pick their own targets. The distinction between controlled and uncontrolled is subtler than you might expect.

In addition, having a human overseeing the decision to release a weapon is only important if you care about collateral damage. Terrorists, insurgents, and believers in total war may not have such scruples.

The other factor to bear in mind is that jamming cuts both ways. Drones are not the only devices that use radios. The counter-drone systems described above are heavily reliant on a radio link between the sensor, the control center, and the weapons platform. Drones that can jam any of the links in that chain will be able to fly safely past without firing a shot.

Get Lost, Drone: GPS Jamming

If blocking communications does not work, we can try another weak point, navigation. A human pilot has many ways of finding their way around, but most drones only have GPS. The power of the Global Positioning System's signal has been compared to a car headlight over ten thousand miles away, making it an easy signal to jam.

However, anti-jamming measures are getting better. Raytheon has developed a sophisticated anti-jam device for GPS called Landshield built around a controlled pattern reception antenna. This has an array of receiving elements that combine to cancel out the jamming from any given direction. When a jamming signal is detected, the array automatically nulls it out. It is like looking through a cardboard tube so you can see a faint light in the distance without being dazzled by lights nearby.

Raytheon's previous generation of anti-jam GPS was the Advanced Digital Antenna Platform, which weighed about ten pounds and was the size of a telephone directory. The new Landshield fits on a silicon chip and may be integrated with military GPS devices, from portable units used by individual soldiers to the GPS-guided Paveway bomb and, of course, drones.

Landshield is currently confined to the military. Improvised drones from terror groups will be easier to jam – for the time being. Other states may have their own version of Landshield sooner or later.

If you can blank out GPS over a wide area it will baffle the swarm, while causing other problems. Originally, GPS was envisaged as a military-only system, but after the shooting down of KAL 007 by Russian air defenses in 1983, the Reagan administration decided to make satellite navigation available to all. In-car SatNav is now ubiquitous. For most people, the days of pulling over to consult a roadmap at every junction are over. Disable GPS, and many people will be lost. That is the least of the problems, as an incident in San Diego proved.

On Monday, 22 January 2007, an electronic warfare exercise being carried out in San Diego harbor accidentally jammed GPS signals across the city. Disruption started almost immediately. The emergency paging system at a hospital stopped functioning. The automated harbor traffic management system stopped working, threatening to throw the port into chaos. Air traffic control at San Diego airport reported problems with their system for tracking incoming aircraft. Some bank ATMs reportedly stopped giving out money.

The reason for this disruption is that many modern systems use the precision time signal from GPS satellites. In some cell phone networks the signal is used to give each mast a unique identity; if it is lost, the mast drops off the network. GPS timing signals time-stamp financial transactions to prevent fraud, and this may be why the cash machines stopped working. Power utilities use the GPS time signal to keep alternating current from different power plants in phase across the grid. If this is lost, then attempts to switch power supplies to channel power to where it is needed become inefficient as the out-of-phase currents clash. This may ultimately produce blackouts.

GPS jamming looks more like a weapon for a swarm attacking an urban target rather than as a way of stopping a swarm.

This vulnerability is one reason why alternatives to GPS are a hot topic. One such is the system developed by Australian company Locata. This uses a network of ground-based 'pseudo-satellites' which give a more accurate fix and would require vastly more powerful jammers to block.

While current small drones are heavily reliant on GPS, that may not be true for much longer. You can already navigate urban areas without GPS thanks to Wi-Fi. Each Wi-Fi hotspot has its own fingerprint, including a Service Set ID and Media Access Control address, and transmits them continuously. Service providers including Google and Navizon map out the location of each node; by identifying those closest to you, you can pinpoint your location within less than a hundred feet.

Other researchers are navigating using "signals of opportunity," including not only Wi-Fi but cell phone signals, radio and television transmitters, and other sources of radio waves. These may not be as accurate as GPS, but they can be used indoors as well as out, and cannot be stopped except by jamming absolutely everything. And while a Zeppelin could be fooled by moving transmission from one radio mast to another, the sheer number of emitters means that such deception is now impossible.

Even jamming all radio signals will not stop future navigation. Smartphone developers are looking at navigating via landmarks and topography. The idea is not a new one. Tomahawk cruise missiles were originally equipped with terrain-matching radar to compare the scenery with an electronic landscape map to determine their location. The difference now is that every smartphone has the storage and processing power to scan the scenery and find out where it is. In 2014 Blue Bear Systems demonstrated a small drone fitted with a system called SmartBoomerang. Developed with the assistance of the University of Bristol, this is capable of Simulataneous Location and Mapping (SLAM), making a map of an area as it is flying through. This map can be passed to other drones so they can navigate through the same area without GPS or any other external assistance.

As we saw in Chapter 6, Qualcomm is working on a SLAM device for smartphones using camera input to build up a map of the surroundings as you go through them. Small drones are likely to acquire this capability shortly. And the miniature LADAR sensors described will make SLAM even simpler.

Jamming GPS may look like an easy solution. In practice, though, while it may stop some of the present generation, it will not be effective for much longer.

Fighting Swarms with Swarms

There may be only one straightforward way of stopping a swarm. Anti-aircraft missiles and radar are, at best, a partial solution at best to defending airspace; ultimately, the best weapon against an aircraft is another aircraft. Similarly, the best defense against a swarm may be another swarm.

The leader in this field is Timothy Chung, an irrepressibly enthusiastic scientist at the Naval Postgraduate School in Monterey. Chung has been working on swarms for years, but has only recently started looking at counter-swarm tactics.

Chung is very aware of the potential for cheap drones, and the fact that several of them may cost less than a single anti-aircraft missile. He is equally doubtful that guns or lasers are the answer. Rather than a cheap missile like Spike, why not use a cheap drone with greater maneuverability and longer range and endurance than a missile?

The project's official name is "A System-of-systems Testbed for Unmanned Systems Swarm versus Swarm Development and Research." Chung prefers the catchier title of "Aerial Combat Swarms."

His aim is to stage a contest in which swarms of small drones carry out simulated battles. The first big landmark will be a fifty-versus-fifty demonstration originally scheduled for 2015 but now pushed back into 2016 at least, in part due to sequestration and regulatory hurdles. The regulatory issues have at least now been largely been overcome. It is likely to be the prelude to far bigger contests.

In the process, Chung is learning a lot about how a defensive swarm would work in practice. For a start, he believes the individuals will have a greater degree of intelligence than birds. The swarm as a whole will be able to carry out a complex mission such as defending an area in a highly flexible and reactive fashion.

The first stage is to build a robot army. With Ravens costing $30,000 a time, and faced with a limited budget, Chung needed something more economical to work with. He identified a self-assembly kit for a suitable, easily-modified commercial drone of similar size. This was the Zephyr II drone, which has a wingspan of about five feet. Early models were assembled by Chung's team but as numbers increased and over a hundred were needed, this was outsourced.

Further cost savings were achieved by the team actively developing their own control hardware and software alongside the vibrant open-source community. As with MITRE, the board decision-making will be provided by a cell phone–type single board computer, and the software leverages open-projects. The costs currently stand at about $2k per aircraft, including parts and labor, cheap by Pentagon standards. Chung believes that the end cost per plane will be hundreds rather than thousands of dollars, as he sees a steady stream of cost-dropping technologies hitting the market.

In 2015 Chung's team launched fifty drones into the air at once, using a homemade, chain-driven launcher, believed to be a new world record for fixed-wing drones. Once the drones were in the air, the swarm control system was successfully tested in operation.

Given fifty drones, the normal practice of having one controller per drone becomes impractical. Onboard pre-programmed behaviors allow drones to navigate using waypoints, so one operator might be able to handle a group of ten craft, although at present this is strenuous for a lone operator. There are other ways of working, though, and Chung suggests that in the future, swarm management may be "decomposed." For example, one team member might be in charge of overall swarm behaviors and their execution, while another monitors the health of the swarm; a third reviews the intelligence data the swarm may gather, while a fourth maintains a mission overview as "mission commander."

In practice, Chung says, the drones will have to make many decisions themselves, such as how to avoid collisions and maneuver in formation, because of the speed at which things move, especially in a combat situation. Deciding which enemy to chase or whether to dodge out of the way requires split-second thought processes better suited for onboard computation.

Another key issue is how communication within the swarm works, or in some cases, does not work. While this is a work in progress, and Chung says continued experimentation is needed to achieve the right balance of simplicity and effectiveness, initial tests using inexpensive, USB-WiFi dongles provide the ability for drones to communicate amongst themselves and share relevant data, such as GPS positions of all teammates in the swarm.

The scenario that Aerial Combat Swarms will enact in the planned fifty-versus-fifty is a straightforward one. Both swarms will have attackers that attempt to reach and land in the opponent's target zone and defenders that will try to stop such incursions by "tagging" them before they reach the zone. While initial experiments will use an automated GPS-based scoring system to record a "hit" when a Blue drone gets close enough to a Red, future visions for the competition call for onboard "laser-tag" style systems to register such "tags."

What air-to-air drone combat would actually be like is an open question. When the numbers start rising, the lessons learned from manned fighter combat may be of limited use and we are in new territory. It is not that the research is secret, more that nobody really knows yet. Key issues, such as the relative importance of quantity and quality, remain very much open.

In a six-versus-six situation, it may be tactically easy enough for each of the defenders to home in on one attacker. With fifty versus fifty, it is another matter. If even small numbers start slipping through, the results might be disastrous.

As well as flying actual drones, Chung's team is also carrying out an extensive program of computer modelling of swarm-versus-swarm combat. One of Chung's collaborators has experience with the computer gaming industry; another specializes in multi-agent pursuit modeling, and yet another is investigating computer vision for onboard detection and tracking of opposing drones. They will help establish the theoretical and research framework for effective swarm attack and defense. This will then be validated – or disproven – by actual aerial combat experiments.

Chung's project has a modest, garage-built quality to it. Does the future of air warfare really look like students playing with multi-colored kites? It is easy to underestimate small drones, and the low-fi hardware used for the demonstration gives a misleading impression to those who want to see thundering jets. But the project's aim of developing effective swarming tactics for small drones may be the key to warfare in the next century.

Other approaches to tackling swarms all have obvious disadvantages. Any large single weapon system, be it gun, missile, or laser, is likely to be overwhelmed. Against opponents such as terrorists who can only field small numbers of drones, this may not be a problem. But against state actors who are willing to put serious money into a drone swarm, it is hard to see any effective counter except another drone swarm.

There may even be some awareness of this in the rarefied upper echelons of the US Air Force.

Speaking at the Association for Unmanned Vehicle Systems International (AUVSI) annual meeting in 2014, John McCurdy, Director for unmanned programs at the Air Force Academy, suggested that an F-22 Raptor might be accompanied by a swarm of small drones to act as long-range scouts – and to be a protective shield against enemy drones.

"How many UAVs [unmanned aerial vehicles] would it take to win against an F-22?" McCurdy asked the audience. "Surely less than it would take to win against an F-22 guarded by a swarm of UAVs."

When your primary air defense aircraft – the world's premier air-superiority fighter – already looks like it needs to be defended against this new threat, it suggests that the swarms have already won.

REFERENCES

1) Brought down a quadrotor with a shotgun

Kircher, T. (n.d.). Hillview man arrested for shooting down drone: Cites right to privacy. (http://bit.ly/1Q3B1rl)

2) In early 2015, the US Secret Service was carrying out flight tests in Washington DC

U.S. Secret Service holds nightly drone tests over Washington, D.C.. *NY Daily News*. (http://nydn.us/1NK0WEs)

3) Combating Terrorism Technical Support Office vs drones

Magnuson, D. (n.d.). Counterterrorism officials look for ways to stop small unmanned aircraft. *National Defense Magazine*. (http://bit.ly/1HJ59G3)

4) London's answer to the Zeppelin raids was a 75mm cannon mounted in the back of an armored truck.

Fegan, T. (2002). *The baby killers: German Air raids on Britain in the First World War*. Barnsley: Pen & Sword.

5) "You have to blow them up, to damage them doesn't mean much."

Kamikaze Attack, 1944. (n.d.). (http://bit.ly/1AoJQPB)

6) Sergeant York problems

Connell, J. (1986). *The New Maginot Line*. London: Coronet Books, 1986.

7) USAF 2014 Posture Statement (http://1.usa.gov/1NsImu6)

8) Black Dart 2014

Faith, R. (n.d.). Inside 'Black Dart,' the US military's war on drones. *Vice News*. (http://bit.ly/1vbMlCu)

9) Counter-UAS

Hambling, D. (n.d.). US Army prepares to face small drone threat. *Popular Mechanics*.

10) "ARDEC's focus is on the development of an affordable close-in counter-UAS system"

Hannibal People Interview with author, February 2014

11) 1973 laser drone shoot-down

A brief history of the Airborne Laser. (n.d.). (http://1.usa.gov/1XD7QA0)

12) GBAD

Morrison, P. Naval solid state laser program overview. (http://bit.ly/1PBiWRF)

13) US Army HELM

High energy laser mobile demonstrator. (n.d.). (http://bit.ly/1LRfw5v)

14) LaWS

Scott, R. (n.d.). Laser weapon breaks cover on USS Ponce. (http://bit.ly/1XL49To)

15) Chinese Counter-UAS laser

China unveils laser drone defence system. *Agence France Presse*.(2014, November 3). (http://bit.ly/1tuB2XR)

16) Adsys counter-laser system

Counter directed energy weapons. (n.d.). (http://bit.ly/1IGPKkk)

17) AESA Radar use as a weapon

Fulghum, D & Barrie, D. (n.d.). Radar becomes a weapon. *Aviation Week*. (http://bit.ly/1QX1o2v)

18) EMP Grenade

Hambling, D. (2009, March). New army weapon aims to fry gadgets, people. *WIRED*. (http://bit.ly/1XL4ghW)

19) CHAMP and super-CHAMP

Peterkin, R. (2013). Selected directed energy research and development for U.S. Air Force aircraft applications: A workshop summary. (http://bit.ly/1QiqiIh)

20) Fritz-X

Bomb, guided, Fritz X. (n.d.). (http://s.si.edu/1Qiqkjj)

21) Warlock Green and related jammers

Warlock Green / Warlock Red. (n.d.). (http://bit.ly/1IGPRfG)

22) Anti-UAV Defence System

Counter drone anti-UAV system unveiled by British trio. *Security Middle East*. (http://bit.ly/1TnUXDs)

23) QR Code signalling

Lucas, A. (n.d.). Digital semaphore: Technical feasibility of QR Code optical signaling for fleet communications. (http://bit.ly/1XL4oxO)

24) Raytheon Landshield

Raytheon launches enhanced Landshield GPS anti-jam capability. (n.d.). (http://rtn.co/1OAIZXI)

25) GPS jamming effects on infrastructure

Hambling, D. (n.d.). GPS chaos: How a $30 box can jam your life. *New Scientist*. (http://bit.ly/1jBDivN)

26) Blue Bear SLAM

Successful SmartBoomerang demo at Blue Bear. (n.d.). (http://bit.ly/1SAubr6)

27) Aerial Combat Swarm

Hambling, D. (n.d.). Swarm defense. *Aviation Week & Space Technology*. (http://bit.ly/1QiqGGO)

28) Chung is learning a lot about how a combat swarm would work in practice

Timothy Chung interview with author, August 2012

29) How many UAVs would it take to win against an F-22

Mehta, Aaron. "Next-Gen UAVs: Armed, Modular and Smaller." (http://bit.ly/1N6e5lm)

CHAPTER TEN

WHAT HAPPENS NEXT?

"Science has not yet mastered prophecy. We predict too much for the next year and yet far too little for the next ten."

- Neil Armstrong

When Nikola Tesla developed a radio-controlled torpedo boat back in 1892, he believed that it could end war. Small powers could fight off larger ones, and aggressive military action would be impossible.

"War will cease to be possible when all the world knows tomorrow that the most feeble of the nations can supply itself immediately with a weapon which will render its coasts secure and its ports impregnable to the assaults of the united armadas of the world," Tesla told the New York Herald. "Battleships will cease to be built, and the mightiest armorclads and the most tremendous artillery afloat will be of no more use than so much scrap iron."

The modern-day prophet of unmanned systems is unlikely to suggest that they will abolish war. The twentieth century provided all too much evidence that no weapon, however effective, could achieve that, though they might reshape the way wars are fought.

We may also harbor doubts about how quickly things will change. Tesla's anticipation that everything would change tomorrow is not realistic. Change is gradual, and Armstrong's remark about predictions one year ahead and ten is accurate when it comes to drones. While the media may prefer dramatic headlines, the swarms will not be conquering the world next year. However, the trend has started, and swarms will advance over the next decade.

In the world of civil drone aviation, prices have plunged in an intensely competitive market. As the FAA fumbles with regulations on commercial drone operations, companies like Amazon and Google are gearing up to offer delivery services with fleets of small unmanned aircraft. These will have drones crisscrossing major cities, delivering packages right to the customer's doorstep or balcony, but only when the flight regulations are ready. In Switzerland, the first postal deliveries by drone have already taken place, and every week there are stories of new drone delivery trials around the world.

A few years ago, eyebrows would have been raised if you suggested that your company buy a drone for aerial photography. Even if the idea had been accepted, you would have been limited to a few established suppliers, and drones were a highly specialized field. Now high-quality drones can be bought online, and plenty of companies offer training in drone piloting. If the job is a straightforward one, anyone can purchase a ready-to-fly drone like the DJI Phantom and take professional-quality video for a few thousand dollars. It is a fraction of the cost of hiring an aircraft or helicopter and may be cheaper than getting a crane in to inspect a chimney or other hard-to-reach site.

The line between consumer, prosumer, semi-professional and fully professional is blurred. As with computers and other electronics, what was the standard for industry professional a few years ago is increasingly available to the average teenager. Drones that take pictures, and even drones that can deliver a payload, are not the province of the few. Chinese drone maker DJI was said to be shipping twenty thousand units a month in 2014, with a sharp uptick at Christmas. The company is now set to hit the billion-dollar mark and, according to some estimates, more than a million drones will be sold in 2015.

So, private citizens and small businesses are already buying cheap drones by the truckload. What about the military?

A Resistance to Swarms

Chapter 4 showed how the military might acquire cheap drones with similar capabilities to Raven scouts and Switchblade strike drones. There is currently little sign of an appetite for this approach in the Pentagon.

As we saw in Chapter 1, the American military has a poor relationship with drones. In fact, it can barely force itself to use the word *drone*, still preferring Remotely Piloted Vehicle, Unmanned (or Uncrewed or even Uninhabited) Aerial Vehicle, or other circumlocutions. The military establishment is equally averse to the idea of swarms, and again speaks a different language of teams and constellations and formations. It is a fair guess that, apart from some advanced thinkers, the policy senior commanders at the Pentagon are not going to embrace drone swarms anytime soon.

The National Defense Authorization Act for Fiscal Year 2001 stated Congress's intention that "within ten years, one-third of US military operational deep strike aircraft will be un-manned," demonstrating that politicians like drones well enough. Needless to say, fifteen years on the number of deep-strike drones remains at zero. Long before this there were suggestions that rather than insisting on expensive, exquisite aircraft – in particular the billion-dollar B-2 bomber -- the Air Force should be buying simple cruise missile carriers, These only had to get to within a thousand miles of the target to launch their missiles; these would not need stealth, agility of complex countermeasures. But the Air Force remained firmly wedded to the idea of expensive planes. The next-generation Long Range Strike Bomber, expected in service by the mid-2020s, is set to be even more pricey than the B-2.

In fact, something like the drone swarm has been proposed before. The Air Force had a plan for "wide-area persistent munition" in the early 2000s. This was known as Dominator because it would dominate an area. It had some of the characteristics of the drone swarm, with kamikaze drones or "persistent munitions" weighing about a hundred pounds with a wingspan of about twelve feet. The wings fold back for transport, leaving a boxy shape eight inches square and four feet long that can easily be stacked in small spaces. A transport aircraft would carry the drones in pallets of twenty, launching them from the air. A single plane could carry thirty pallets, releasing six hundred drones to blanket the area.

The Dominator drones, made by Boeing, would circle inside enemy territory. Though not especially fast or stealthy, their sheer number would make them impossible to shoot down by normal means. The drones would provide "Just In Time Strike Augmentation," a concept taking its name from "Just In Time" business methodology, which aims to achieve efficiency by delivering things exactly when they are needed. The mass of Dominators in an area would ensure that there was always one close by when a target presented itself, so there would be no delay in between spotting a target and engaging it.

The first Dominator prototypes flew in 2006. The drones themselves worked well enough. Each Dominator carries three warheads capable of destroying a tank, truck, parked plane or other such targets. The collective effect of a number of Dominators would be far more devastating than Reapers or other single, large drones. But the program has not sparked any detectable enthusiasm, and appears to be surplus to requirements. There is not perceived to be a gap for it to fill.

Technically, the Dominator program is still alive, but at the time of writing, the next phase is not expected to be completed until 2017. Given the current budget situation, it is not certain that it will even survive that long. While it may emerge and fly in some form, the Dominator seems more likely to be an air-dropped reconnaissance and light-strike platform deployed by ones and twos. It might be sent in where the threat level is too high for manned aircraft and the Air Force does not want to risk a Reaper. Dominator could fill a niche somewhere between long-endurance Predator drones and traditional smart bombs, making it a useful capability but not a revolutionary one.

The US Navy has also looked at swarming unmanned systems. The KILSW program was started in 2006. CICADA is a glider and might be loosely described as a paper plane made out of a circuit board. It has a wingspan of a few inches and like the Dominator it is designed to be stackable. Eighteen CICADAs can be packed together in a six-inch cube. Like the Dominator it would be dropped from the air to cover a wide area.

CICADAs are designed to be self-deploying sensors. On landing, they would network together like the sensors in SPAN, tracking anything that moves through their zone of operations. In addition to sensors, CICADAs could also carry what the developers refer to as "effectors," probably a euphemism for high-explosive warheads. Unlike a minefield, CICADA would be able to intelligently monitor activity and request approval from a human operator before destroying anything, so it would bypass international laws against landmines.

Like Dominator, the CICADA is not exactly dead, but there is little sign of life. In 2012, two CICADAs were dropped from a drone mothership. They successfully glided to a destination eleven miles away, landing within fifteen feet of the target. CICADA surfaced again in a small demonstration in 2015. Again, there is little indication that this is the precursor to a major effort, but is being kept alive by a small band of believers.

In the early 2000s, the Naval Surface Warfare Center carried out a project called Smart Warfighting Array of Reconfigurable Modules or SWARM, with ten small drones that communicated directly with each other. In 2005, Alion Science & Technology Corp in Chicago received a $20 million contract to develop software for swarming drones. The plans was for multiple drones to work autonomously and act together with cooperative behaviors, but nothing seems to have come of it.

The situation may be changing. As mentioned previously, in April 2015 the Navy unveiled a new project called LOCUST – contrived from Low-Cost UAV Swarming Technology – which aims to get thirty small drones into the air and flying together in 2016. LOCUST is a small start, but it does suggest that some within the Navy recognize the offensive and defensive potential of the swarm. Whether it fares better than Dominator or CICADA is another matter.

When it comes to fielding small, powerful drones, the Air Force may have the lead. The USAF's Micro-Munitions program promises to take precision strike to a new level. It is an air-delivered drone the size of the Raven or Switchblade with folding wings. It will be able to operate for weeks on end thanks to a combination of perching, power scavenging, and solar cells. The fixed-wing drone will be able to navigate urban canyons and to target individuals, a degree of finesse impossible with existing drone strikes. If used in significant numbers, Micro-Munitions may give the first indication of what a drone swarm can do in action. However, as with other programs the timescale has slipped from the initial goal of 2015 and it is not clear when the Micro Munitions will be fielded.

Like Switchblade, Micro Munitions may only be launched in small numbers against specific high-value targets. Without the sort of swarm control developed for LOCUST, the manpower (and bandwidth) demands of piloting large numbers of drones are unmanageable.

It is significant that the USAF are casting small drones as munitions rather than aircraft. The glory will still go to the F-35 that swoops in and delivers the Micro-Munitions to the target area, even if the job could be done by a cargo aircraft. Politics requires that the drones supplement the fleet, not supplant it. As with the F-22 defensive swarm idea, the drones feature as auxiliaries only.

So, although the technology has been floated past them several times, the Pentagon is nibbling at the edges of the idea without really biting. There are no voices prophesying doom for existing aircraft, warning that the advance of technology puts manned jets in the place of battleships in the 1920s. Small combat drones are still seen as irrelevant to the bigger strategic picture, even if they do have their tactical uses.

Nor is there much encouragement for small drones from the military aviation industry. It would be suicidal for a sector that earns hundreds of billions of dollars of revenue from programs like the F-35 to turn its back on manned aircraft. The mass production of small drones cannot generate the profits needed to keep the industry going. Apple makes money from the big difference between the cost of manufacturing an iPhone in China and the amount they can sell it for in the US, hardly an option open to aerospace giants.

Small drones do not require the same sort of manufacturing base and the long secondary tail of specialists in engines, turbine blades, fuel-flow valves, radar components, and the ten thousand other elements that a modern plane depends on. A small drone industry could be outsourced to low-cost producers in the Far East.

The aviation industry continues to put its message across to the public, sometimes via full-page adverts in the newspapers, that we need the F-35 and its kind as never before. They will continue to lobby politicians to keep spending and not be tempted to cut or divert funds. They will advise against any hasty changes to a program that has been going since 1998, and which is now beginning to bear fruit. On current plans, the F-35 production lines need to keep going for another twenty years.

We can expect that any study of small drone swarms from within the industry will highlight their drawbacks and limitations compared to existing platforms rather than trumpeting their potential. This is not cynical self-interest, but a well-established effect known to psychologists as in-group bias. The refrain of "Not Invented Here" used to disparage foreign products or inventions has long been an obstacle in the industry; when faced with something as alien as drones, it becomes the standard response.

Of course, the anti-drone viewpoint is valid. Small drones are not a direct substitute for manned planes; they do not fly as fast or as high or carry the same weapons. It is also true that an aircraft carrier is not a substitute for a battleship, and battleships can do many things that aircraft carriers cannot. Equally, horses can do things that tanks cannot, and the machine gun can never be a true replacement for the bayonet. But when it comes to combat, machine guns trump bayonets, tanks trump horses, and carriers trump battleships.

And the evidence suggests that drone swarms will soon trump manned aircraft.

From Boots on the Ground to Drones on the Roof

While the US and other Western powers are not rushing to embrace drone swarms, a political imperative that may drive them in that direction. It is precisely the same factor that gave drones their foot in the door as spy planes to start with: the unmanned option lowers the political risk to an acceptable level.

Drones were originally seen as useful because another Gary Powers incident would be unacceptable. The same applies today. The loss of a manned aircraft causes serious concern, but when a US drone crashes in Iran or some other forbidden spot there is no great consequence.

In late 2015 there has been much argument over the need for "boots on the ground." Are airstrikes enough, or is it necessary to intervene by sending in ground troops to Syria or other troublespots? The drone swarm may offer an alternative; it has the advantages of direct military intervention without the politically distasteful possibility of soldiers coming back in body bags.

This is a return to the idea of air power as a decisive force. Since its inception, advocates have argued that wars could be won by air power alone. History has failed to bear this out, and their critics have scorned the idea of victory from the air. When it comes to unrestrained destruction, air power can level cities and raze infrastructure. This much was amply demonstrated in Berlin and Tokyo in WWII, not to mention Hiroshima and Nagasaki. Air power can, as Air Force General Curtis LeMay memorably put it, "bomb them back to the Stone Age." But that does not mean it can win.

In particular, air power is less effective for more limited forms of warfare. In Vietnam, a major bombing campaign failed to stop the Viet Cong from building up and supplying their forces along the Ho Chi Minh Trail. In a blunt secret assessment, President Nixon said that ten years of bombing in Vietnam and Laos achieved "zilch". This was in spite of having dropped over seven million tons of bombs, compared to two million tons of bombs in WWII. Proponents point out that later on, the final Linebacker raids helped bring the North Vietnamese to the negotiating table, but this could hardly be called winning the war.

More recently, military analysts have chewed over the lessons of Kosovo and Libya, where aircraft from the US and its allies, along with local ground forces, produced the broadly the desired outcome in the short term. However, both cases exposed the limitations of air power. While air strikes can easily take out military vehicles and government buildings, planners can run out of targets while leaving enemy forces still firmly in control. So long as vehicles remain camouflaged or hidden in bunkers, and soldiers keep their heads down, they are immune to air strikes.

This is even more applicable in an insurgency, when the enemy can attack and then melt away into the hills and mountains, or into cities and villages where they blend in with the locals. Foes like ISIS have little infrastructure and no massed vehicles that can be bombed.

More seriously, air power cannot hold ground. Air forces may be able to destroy or drive back the enemy, but only on a temporary basis. Clear the region of enemy troops on Monday, and on Tuesday they walk right back in. When ground troops occupy a piece of ground, they hold it for good.

This ability to bring the enemy to battle, to take and hold ground, underlies the debate around "boots on the ground," allied foot soldiers who can force the enemy surrender or fight. But while foot soldiers may be effective at getting results, they are politically risky.

Sending the troops in is inherently dangerous, and the public are averse to casualties.

Manned air strikes have some risk of losing a pilot, but ground operations carry an almost inevitable consequence of flag-draped coffins returning. In political terms, it is easier to carry out a hundred air strikes than have one soldier killed.

This leads to costly and elaborate "force protection" measures in which the bulk of the soldier's effort is focused on protecting themselves rather than carrying out their mission. Under these conditions soldiers cannot operate from tents but need fortified bases with armed sentries on duty round the clock. Supply convoys have to be escorted by armed and armored vehicles. Sometimes bases have to be supplied by air because the roads are too dangerous. A smaller proportion of the available fighting power can be devoted to the mission of fighting the enemy, so more forces need to be in-country to carry it out.

This leads to awkward compromise solutions, such as were seen in Kosovo, Libya, and more recently in Syria and Iraq, where massive US air power acted in support of modest local forces. The hope is that undertrained and underequipped allies will provide the necessary ground element, and the outcome is rarely satisfactory.

It is difficult to coordinate a foreign air force with under organized ground troops who lack the necessary communication network. The problem is particularly acute with precision munitions like laser-guided bombs. One solution is to send in a handful of Special Forces to supplement local allies. They can fluently and competently direct air strikes, and use a laser designator to highlight the one building that contains gunmen in the otherwise friendly village.

Unfortunately, even a small number of boots on the ground may be politically awkward. The presence of US forces may be highly counterproductive because it can attract international jihadis who simply wish to "fight the Americans."

Domestic US politics demands that US casualties are minimized. Witness the continuing furor over the deaths in 2012 of two Americans at the diplomatic compound in Benghazi at the hands of Islamic militants. Some nine thousand Libyans have died in the Libyan civil war, where the US effort included strikes by B-2 bombers dropping two-thousand-pound bombs, but these are of less political significance in the US. This sensitivity to casualties means the military can find their mission becomes one of returning home safely rather than pursuing the original objective.

This is why in March 2015, the US withdrew its military forces from Yemen in the face of a growing insurgency from the Iranian-backed Houthi faction. The soldiers withdrawn included Special Forces units who were training the Yemenis in counterinsurgency. These Special Forces had knowledge of the country and the language and might have been expected to spearhead a US military response. In the event, it was deemed politic to get all US personnel out of the area. Thus even the most capable fighters in the world are seen as a liability in a war situation.

Fear of casualties can hamstring US military power. The US Air Force is arguably the most potent military machine in the world; the US Navy's carrier battlegroups can reach every corner of the globe with overwhelming force. But without boots on the ground they may have trouble fighting even modest enemies like ISIS, who can only field a few thousand fighters in pickup trucks.

The process of force projection changes with drone swarms. While an air strike comes and goes – and the effects are debated afterwards – a swarm has the persistence to occupy an area. The swarm can stay in place for weeks or months and get up close. To drone operators, people are not the tiny, ant-like dots far below that the aviator sees, with friend and foe impossible to distinguish. They are close enough to literally recognize faces, and operators can pick out the armed insurgents from innocent farmers.

Small drones promise genuine precision strikes. Rather than the indiscriminate slaughter of five-hundred-pound bombs falling on wedding parties, small drones can single out the one wedding guest known to be a terrorist leader, follow and target him individually at an appropriate time. They can sit beside roads or at the entrance to villages, reading license plates and scanning faces. They can perch outside buildings and wait as long as needed, or search at close range for hidden vehicles or concealed cave entrances. The enemy can be exterminated one man at a time if necessary. If overwhelming force is needed, drone swarms are also capable of destroying entire convoys or building complexes.

Of course, it remains to be seen whether such a promise could ever be fulfilled. Even if the right targets are struck, there will be plenty of claims that innocent people were killed, so it may be hard to judge the truth. Of course, human rights advocates will be concerned that innocent people really are being killed, and with no friendly forces in the area it will be even more difficult for them to investigate.

In any case, the persistent drone swarm can hit an indefinite number of targets, rather than only being able to target a few selected individuals like current drone operations. To anyone identified as the enemy, entering the area controlled by a drone swarm is like walking into a minefield. Stepping over that invisible line means you will be attacked by an endless stream of coordinated, relentless, and merciless drones.

In principle, this capability could largely supersede the need for boots on the ground in some situations. Perhaps the first prototype of this sort of operation has already occurred – Israel's "Pillar of Defence" in Gaza in 2012 was the first major offensive in that area conducted without any soldiers on the ground, in which precision strikes were made possible by a large force of scout and attack drones.

Drones cannot do the all other things that soldiers do besides fighting. In particular they will not be winning hearts and minds by forging friendly relationships with local people. They will not be playing soccer matches, building schools or digging wells. A drone swarm is about as friendly as a set of CCTV masts with machine guns pointed down at the population.

Task Force ODIN, with their fleet of drones spying on the population to catch insurgent bombers, was a small foretaste of what the technology could accomplish. A drone swarm will cover a wider area and watch in greater detail over a longer period. The vast majority of the surveillance will probably be completely autonomous. The limitations of bandwidth and the need for human operators make this inevitable. But as soon as the drones spot anything suspicious, their human handlers will be alerted by a request for instructions.

Future drones may have two-way audio communications so the operator can talk to a potential target. Swarm Systems Ltd has already tested a drone for the British Army with this capability. So future drone strikes may not be a matter of a Hellfire missile striking a truck out of the blue. It might start with a polite request from a hovering drone to lay down weapons and step forward to be identified. Back in 1991, the sight of Iraqis surrendering to a drone was comical. In the near future, surrendering to a small drone may be common sense, given that it is likely to be part of a swarm.

Of course, weapon systems rarely match up to their promise. Suicide bombers can still lurk when there is a drone perched on every roof, and targeting errors will still lead to unnecessary deaths. A drone swarm does not automatically make for quick, surgical wars in which only incorrigible enemies are killed. But the possibility of being able to wage full-scale wars without risking friendly casualties may be politically attractive. The military may not like drone swarms, but a swarm can take the fight to the enemy when there are no other alternatives.

There are plenty of historical precedents. When airpower was the only way for the Allies to strike at Germany during WWII, airpower was used, in spite of the number of civilian casualties it would cause. When cruise missiles were the only way for the US to exert power, they were used – for example against targets in Afghanistan and the Sudan in 1998.

For this reason alone the drone swarm may thrive, in spite of all opposition from within the military, just as the Predator did a generation ago.

Talkin' 'Bout a Revolution

A shift away from a small number of manned aircraft to swarms of drones might be a gradual matter of substituting one weapon system for another. There are already signs that swarms like LOCUST will acts as auxiliaries, expendable scouting and strike assets, accompanying and assisting manned aircraft. They might, for example, be tasked with the dangerous business of "defense suppression," knocking out air defenses so manned aircraft can fly safely. Manned aircraft may increasingly launch small smart weapons that look like smart, armed drones.

However, once the genie is out of the bottle, there may be a quicker and more dramatic shift. Tesla's unmanned torpedo boat did not make existing battleships as useless as scrap iron, but something else did.

When Britain's Royal Navy launched HMS Dreadnought in 1906, she was intended to cement that country's place as the world's leading naval power. Britain already had more battleships than anyone else, and the Dreadnought was essentially an upgraded version of the same thing. Instead, the new design destabilized the naval world, setting off an arms race that helped start the First World War.

Dreadnought was not the first armored battleship, but two features made her unique. Rather than a mixture of large and small guns, it mounted ten twelve-inch guns, giving her superior long-range firepower. Other ships the same size only had four big guns, so in a long-range duel, Dreadnought would win every time. In addition, Dreadnought was fitted with the latest steam turbines, giving a speed of 21 knots (24 mph), compared to the 18 knots (21 mph) of earlier battleships. This small but significant difference meant Dreadnought could escape from any anything, but nothing could escape Dreadnought. Fights would always be on the Dreadnought's terms.

Dreadnought made earlier battleships obsolete – including all the British ones built up over decades. It wiped out more than a century's advantage and gave other powers a chance in the naval arms race. The German admirals realized that the inferiority of their current fleet had become irrelevant. If they could build more Dreadnought-class ships than the British, they would have the balance of power at sea. Other nations, notably the US and Japan, scrambled to build their own.

At present, the US is the world's preeminent military power and reigns supreme in terms of combat aircraft, aircraft carrier groups and or nuclear submarines. The US uniquely maintains the capacity to fight two major wars simultaneously and win both. While Russia and China may be described as near-peers, in terms of hardware they are a long way behind. The US has almost fourteen thousand military aircraft; Russia has less than three and a half thousand, China less than three thousand. When wars are a matter of air power, the US keeps the advantage.

Drone swarms could change that equation. The US's phenomenal military spending – more than a third of the world's total – would ensure more and bigger swarms than anyone else or even everyone else, if some of this money were diverted to buying drones. But it is doubtful whether the Pentagon, with its historic aversion to unmanned systems, would be willing to give up its impressive carriers, glamorous F-35 fighter jets, and awesome M-1 Abrams battle tanks for millions of toy aircraft.

Other nations, however, which may not have the same traditions and entrenched values, and whose economies may be well suited to producing large amounts of small electronics, may see the situation differently.

The iPhone is famously "designed by Apple in California" but largely made in China. The iPhone 6 reportedly costs around $200 to make but retails at over $600. According to some estimates, electronics manufacturer Foxconn, which makes iPhones and iPads, employs over four hundred thousand people in China. This dwarfs Apple's own US workforce of around fifty thousand, and gives an idea of the real size of the industry.

If the Chinese military wanted to build drone swarms, they have the infrastructure to do so in large numbers and at low cost. The $2,000 Razor might start to look as overpriced as the $4,000 laptops of the 90s look now, with tablets with a much higher spec are less than $40 retail. The drones made by DJI cost around a thousand dollars at present, but they are not yet manufactured in the same volumes.

Hundreds of thousands of drones for the cost of one combat aircraft seems feasible, but how much further might it go? Computers are still getting cheaper: the Raspberry Pi Zero sells for just $5. If a basic swarming drone can be built for $200, then it may be possible to buy a million for the price of a frontline combat aircraft. The impact of a swarm of a million drones is as far beyond speculation as the impact of pocket supercomputers was a few decades ago.

Even an impoverished nation like North Korea could produce drone swarms. They may not have much of an electronics industry, but the components are available on the open market and drones could be assembled locally by cheap labor. North Korea could build a weapon of mass destruction capable of reaching the US that cannot be shot down as easily as a ballistic missile. Such a scenario might seem far-fetched, but it would be no more than a modern version of the Fu-Go balloon bombs sent on the same trajectory seventy years ago. Like the Fu-Go, small drones could slip through existing defenses. Unlike the Fu-Go, they would not be dropping bombs at random but attacking exact targets. In 1945, one Fu-Go hit a strategic target when it cut off power to the Hanford plutonium plant; next time they might all hot targets equally as important.

Aircraft change warfare. In the First World War, people in London were shocked that the capital of a great empire should be so open to attack from the air, that there was no way of stopping the Zeppelins. In the 1920s it was increasingly believed that "the bomber will always get through" and that civilian populations would be attacked ruthlessly.

Since the Second World War, the people of Western Europe and the US have been happily insulated from the effects of conflict. We no longer look to our air-raid shelters and gas masks every time our armies are sent to a foreign war. Enemy drone swarms could bring the threat right home, as the people that we are targeting with drone strikes turn around and target us back.

The Build-It-Yourself Air Force

Not everyone has the Pentagon's luxury of picking and choosing what sort of arsenal they wish to purchase. When money is tight, cheap and effective systems rise to the surface.

There has been conflict in the east of Ukraine since 2014. Militias in Russian-speaking eastern Ukraine, apparently with covert backing from Russia, have been fighting against government forces. The Ukrainian Ministry of Defense has not always been efficient in providing the best equipment; critics say that it the procurement process is managed badly and undermined by corruption. Ukrainian forces needed drones for reconnaissance and surveillance on the front line. When they could not get them through official channels, they took procurement into their own hands.

The Aerorozvidka project and the People's Project are both devoted to making up the Ukrainian drone shortfall. Using private money – and in the case of the People's Project, crowdfunding – they have adapted commercial drones for battlefield operations. At the larger end of the scale, the PD-1 "People's Drone" is a fixed-wing craft with a ten-foot wingspan and a pusher propeller. Many of the other drones are simply commercial models like the DJI Phantom supplied by supporters in the US.

This sort of machine takes near-Hollywood quality aerial footage with a stabilized 4K camera for less than $4,000. Dual controllers mean that the workload can be split in the same way as the Predator and Raven, with one operator concentrating on flying the drone while the other aims the camera.

The Ukrainians fit their drones with improved communications, as their opponents have access to effective Russian jammers and have jammed both drone communications and GPS navigation signals. They have also learned to be careful about placement of communication antennas when controlling the drones, and the need for controllers to keep moving. On two occasions, their drone operators have been hit by mortar fire when the Russians pinpointed them by their radio signals.

The Ukrainian drones can find and track the enemy and direct artillery fire in real time. Aerorozvidka even claims to be working on drones armed with homemade missiles.

Any Western soldier who has to expose themselves to enemy fire during reconnaissance may well wonder why they do not have access to the same small drones as the Ukrainians.

Other armed groups that are not part of an established government lack the infrastructure to operate an air force but have used small drones to some effect.

Hezbollah, an Islamist group based in Lebanon and supported by Iran, has been flying small drones since 2004. They have a number of fixed-wing drones called Mirsad with a ten-foot wingspan, which they claim to have developed themselves. These appear to be modified from Iranian drones.

The Hezbollah drones have carried out spy missions over the border into Israel, as well as unsuccessful attacks. The Mirsad can carry up to eighty pounds of explosive, but the slow-flying drones have been picked off by Israel's extremely efficient air defenses before reaching their target.

Similarly, the Palestinian militant group Hamas has flown small drones, also apparently derived from Iranian models, and claims that some of them carry explosive warheads. In 2012 the IDF reportedly destroyed a Hamas drone-making facility; a year later Palestinian police arrested activists building "suicide drones." In 2015 Hamas drones from the Gaza Strip allegedly flew over Egyptian airspace. Under terms of the 1979 peace agreement with Israel, the Egyptians are not allowed anti-aircraft weapons in the Sinai so could not stop them.

The Israelis have highly effective air defenses, and drones like these will not overtax them. Air defense systems like Iron Dome have proven effective in shooting down the small rockets fired by Hamas and other groups. Rockets are small, fast, and more difficult targets, and drones at any altitude are easier. However, again it is a question of numbers. Small, smart drones at treetop height or below might be difficult to stop.

Meanwhile in Syria and Iraq, the al-Qaeda offshoot known as Islamic State or ISIS is also using drones. While previous Islamic extremist groups have largely shunned technology, ISIS has been notable for its adept use of social media and other modern developments, including drones.

ISIS showed footage taken by DJI Phantom drones in August 2014 for their propaganda videos. In December 2014, during the battle for Kobane, ISIS uploaded footage of suicide bomb attacks in the city taken from the same drones.

Like the Ukrainians, ISIS then reportedly started using the Phatoms for tactical reconnaissance, before upgrading to small fixed-wing drones. In December 2015, fighters from the Kurdish YPG reported that ISIS were using hand-launched drones packed with explosives, the type tentatively identified as a Skywalker X8 hobbyist drone. These first attacks were unsuccessful, but they may mark the start of a new era in terrorist attacks.

Drones From Below

Activists or terrorists use of drones will start, like the Ukrainians and Timothy Chung's drone air force, with simple modifications to existing drones. However, any larger-scale effort is likely to follow the model set by MITRE with their Razor drone.

We can get some idea of how this might progress from a crowdsourced South African project to help with conservation: the "Wildlife Conservation unmanned aerial vehicle Challenge" or wcUAVc. This is an initiative to protect endangered species and stop poaching and wildlife trafficking using small drones. The challenge is to come up with low-cost drones that can be deployed by rangers in the Kruger National Park in South Africa. As the wcUAVc puts it:

"UAVs that governments use in warfare might provide some of these capabilities, but Kruger does not have the runways, crews, maintenance facilities, and billions of dollars needed for such operations. The challenge is to design aircraft that can be launched in the bush, operate for hours over the rugged terrain, detect and locate poachers, communicate over existing commercial infrastructures, and recover in the bush – all for under $3,000."

Over a hundred teams from around the world are taking up the challenge, mainly universities and groups of students. The goals include endurance of up to two hours, a range of twelve miles, and being able to detect, locate, and identify humans and animals from an altitude of several hundred feet in dim lighting. Like a military drone, the anti-poaching version needs to be able to tell whether people are carrying weapons. The wcUAVC recommends a specific cheap ($109) video camera for this task.

As with MITRE, the wcUAVc effort focuses on 3-D printing and commercial off-the-shelf components to build effective low-cost drones. As the effort progresses it may produce techniques and software, such as autopilot and object-recognition algorithms, that will be shared openly. The military and industry may not bring about the revolution in small, low-cost drones. It may come from below.

There have been a few homebrew drones like the Wasp, the hacker's tool to sniff out Wi-Fi and 3G and aid in breaking computer security described in Chapter 4. This may be a foretaste of the future. If the Ukrainians can crowdsource their military drone procurement process, so can others.

"Terror Drones"

In the past there has been little sign of a real drone threat from terrorists. There have been numerous "drone terror plots" and convictions, but none of the cases have amounted to very much.

A well-known example is 26-year-old Rezwan Ferdaus, sentenced in 2012 to seventeen years in prison for plotting an attack on Washington. His plan included three radio-controlled model planes crashing into the Pentagon and the Capitol carrying five pounds of plastic explosives each. These would have been followed by a ground assault by Ferdaus and others he somehow planned to recruit.

Ferdaus, who is a US citizen, played drums in a rock band before becoming radicalized and deciding to wage his own personal jihad. His drones were literally toys, model F-86 fighter jets. Ferdaus had already assembled one when he was arrested. It is a crude device; the radio link can be easily jammed, and with no navigation or camera on board, it relies on the distant operator sending it in the right direction. How accurately it could be aimed is open to question. It is unlikely the attack would have done much damage, but apparently Ferdaus was relying largely on the psychological or perhaps symbolic impact of drones attacking Americans.

This is typical of the sort of drone terror stories so far, being based largely on fantasy. An extra air of unreality was added when it emerged that Ferdaus was the victim of an FBI sting, in which two agents posed as al-Qaeda terrorists. They encouraged him to produce a detailed attack plan, supplied him with weapons and explosives, and gave him several thousand dollars to buy the radio-controlled planes. These actions have led to allegations of entrapment, and it is hard to believe that Ferdaus' dream of jihad would have come to anything without FBI assistance.

Two factors have prevented drones from being a serious terror threat to date. One is that delivering a bomb with a drone is inherently more complicated than a suicide bombing or a gun attack. It requires several additional steps, each of which brings the possibility of being caught, and requires a higher degree of technical skill. Flying model aircraft is something that takes a hobbyists patience and skill. Experienced terrorists prefer simpler plans that are more likely to succeed.

Secondly, much of the current terror threat comes from Islamist groups who favor suicide attacks that they portray as martyrdom. A drone may be seen as cowardly or, worse, making them morally equivalent to the Americans. It is hard to whip up condemnation of an enemy because they use drone strikes when your own side employs the same tactic.

The future threat may be different. Modern drones are designed to be flown out of the box by an inexperienced operator. Groups like ISIS have shown themselves more adept with technology than previous Islamic extremists. In addition, groups like Anonymous, which are more oriented towards high-tech direct action, currently do considerable damage by hacking and cyber attacks. Small drones would be a natural extension of their operations.

In April 2015, Japanese citizen Yasuo Yamamoto was arrested for landing a drone on the Prime Minister's official residence in Tokyo. Yamamoto turned himself in at a police station. The drone was marked with a radioactive hazard warning sign, and Yamamoto says the drone was carrying radioactive sand from the Fukushima nuclear power station site, his protest against nuclear power. It was a simple but effective protest, and pictures of officials cautiously removing the radioactive drone went around the world.

Protests are unlikely to involve lethal attacks but may be aimed at disrupting of organizations that have merited the attacker's disapproval. For example, environmental extremists may wish to shut down a polluting power station, halt construction work in an ecologically sensitive area, or prevent a new airport from operating. Drones cannot be stopped by chain-link fences and security guards.

Paint bombs and spray cans are a highly visible way of protesting, and with a drone the entire world is your canvas. In April 2015 a drone scribbled a tag across the face of a model on one of New York City's biggest and most viewed billboards. Video of the event, staged by an artist called KATSU, went viral. The drone was a DJI Phantom, modified to carry an aerosol spray can. Later in 2015 Mexican activists deployed their own modified Phantom called "Droncita" to deface posters of their President with red paint.

Such interventions are likely to grab headlines and highlight how difficult it is to stop drones. They may pave the way for more alarming uses.

Drones look like horribly effective terror weapons. They can be launched from miles away and fly over the security barriers that stop car bombs, suicide bombers, and others from reaching a target. They are difficult to trace. And they are highly visible: simply by swooping over crowds in a stadium or loitering in an airport approach path, drones can make news headlines and create a sense of panic.

Terror drones have the benefits of precise placement and timing. Good security makes it difficult for would-be assassins to get close to a political leader, but drones change that. In Germany in 2014, the Pirate Party crashed a Parrot quadrotor onto the podium in front of Chancellor Angela Merkel during a political meeting. It was a publicity stunt, a protest against the EU's use of surveillance drones, and again it made headlines. The next drone to crash next to a politician might be carrying an explosive warhead.

Unlike the largely imaginary threat of printed guns, printed drones could represent an entirely new sort of danger. They are a distributed threat: rather than relying on central control, an attack could be crowdsourced. Independent cells could all build drones to the same simple plans shared online, and release them in a given city at a given time without ever meeting.

This type of crowdsourced drone would not have the sophisticated weaponry discussed in Chapter 8. Even with improvised explosives or incendiaries, small drones could do massive damage. They could distribute improvised chemical weapons, like the home-brewed Sarin nerve gas used in the Tokyo subway attack.

More importantly, they would spread fear. Terrorism is about creating terror in a population. A highly visible attack that the authorities are helpless to stop might provoke a response as dramatic as 9/11, even if it only caused a fraction of the casualties. There is no video footage of the first plane striking the tower on 9/11, but any terrorist group using drones could capture plenty of drone video footage and post it online. Players like ISIS have already mastered both the use of drone filming and online distribution and this would magnify the media impact of any strike.

The Future

Drone swarms are likely to be a defining weapon in future wars. It is impossible to tell what will happen in conflicts between two powers which both use them. The simplistic view would be that the bigger or better swarm wins and destroys the other utterly. But it is equally possible that, as with air warfare in WWII, there is no overall dominance. In 1940 Britain and German were able to bomb each other. Given the destructive capabilities outlined above, the effects of even part of a swarm could be apocalyptic.

As we have seen, the drone swarm increases the political temptation to intervene in foreign wars. But what happens when two or more outside powers send in their drone swarms? Escalation looks increasingly likely. The swarms' ability to pass over borders may also tend to spread regional conflicts into wide areas. Given that deniability remains a key feature of drones, then spying or sabotage by 'black' swarms which are not acknowledged by their controllers are also a possibility. One can easily imagine an angry Ambassador waving a damaged drone in the UN Assembly, insisting that it came from a neighboring country, while the opposing ambassador looks on with a sneer of contempt at this obvious lie.

Again, the nuclear balance is maintained because neither side can disable the other's strategic weapons with a first strike. Swarms might change this balance and make first strikes possible – or strikes by non-nuclear powers seeking to disarm nuclear ones. The swarm club is far easier to join than the nuclear club or the ballistic missile club.

A swarm might be deliberately engineered as a weapon of mass destruction, with no mechanism to recall or halt it. While few dictators are able to get their hands on atomic warheads and ballistic missiles, swarms are more easily accessible, and the threat of a "doomsday swarm" designed to kill as many people as possible might be a counter to nuclear-armed opponents.

Such possibilities indicate that swarms could be highly destabilizing. That might encourage the international community to start discussions on arms control and whether swarms should be classed as weapons of mass destruction.

Then there are the drones themselves. There has been much concern over the future of artificial intelligence. The media express this in terms of robots running amok. This taps into a basic human fear of our own creations, a fear with a rich cultural history from *The Sorcerer's Apprentice* via tales of the Golem, *Frankenstein* and Karel Capek's *RUR* to the *Terminator* series and a thousand other movies.

The media tend to over-dramatize, but there is concern among genuine experts in artificial intelligence. As already discussed, the type of drones discussed here are unlikely to be given a high degree of independence, with the US military emphatically committed to keeping drones under human control and limiting their autonomy. While the dangers of artificial intelligence should not be underestimated, military hardware is likely to be kept under much tighter control than any other robots.

What ought to be far more alarming is not that drones will refuse to obey orders, but they *will* obey them.

Historically, dictators founder at the point where their troops refuse to open fire on their own people. The Russian 1991 coup failed when troops would not storm the government building in Moscow known as the White House after it was taken over by protestors. With drone armies, there is no risk of such disobedience. Drones will attack whatever they are told to attack, whether it is a command bunker or a children's hospital.

From a practical perspective, problems with bandwidth, communications jamming, and the sheer volume of target data that a swarm can generate, mean that some swarms may be treated like cruise missiles. They will be pointed at a target and sent on their way.

There may be legal challenges to drone development. As with cluster bombs and land mines – both weapons that are so dangerous because they are fielded in vast numbers – the drone swarm may be outlawed to some degree. Organizations like ICRAC, the International Committee for Robot Arms Control, have campaigned tirelessly to raise public awareness of unmanned weapons and to limit and regulate the use of armed drones. So far, they have had little effect, but as awareness of the new technology increases and the potential threat becomes more obvious, many more voices may join them in the coming years.

In the mean time we will see how small drones evolve and who decides to take advantage of the opportunities they offer. Because of the way the defense aerospace establishment works, the US and other Western forces start at a disadvantage. The next ten years may see more of a shift in military power than anyone has been expecting.

There is little doubt that swarms of small drones can conquer the world. The question is who will be controlling those swarms, and who will be under their control.

REFERENCES

1) DJI Inspire

DJI official product page "Inspire 1" (http://bit.ly/1xQxWhL)

2) Boeing Dominator

Hambling, David. "Drone Swarm for Maximum Harm." *Defensetch.org.* (http://bit.ly/1OAJocB)

3) Close In Covert Autonomous Disposable Aircraft (CICADA)

"Close-In Covert Autonomous Disposable Aircraft (CICADA) for Homeland Security." *Mistral Solutions.* (http://bit.ly/1PBkuuO)

4) F-35 advertising campaign

Thomson, Mark. "Child Warfare: Dogfighting Brothers." *TIME.* (http://ti.me/1XL4Fkp)

5) Ukrainian drone projects

Tucker, Patrick. "In Ukraine, Tomorrow's Drone War Is Alive Today." *DefenseOne.* (http://bit.ly/1HuqWwc)

6) Hezbollah Mirsad drone

Hoenig, Milton. "Hezbollah and the Use of Drones as a Weapon of Terrorism." *FAS.ORG.* (http://bit.ly/21zjBXx)

7) Hamas drones

Winer, Stuart. "Hamas drones Said to Enter Egyptian Airspace."

8) ISIS drone

Tadjdeh, Yasmin. "Islamic State Militants in Syria Now Have Drone Capabilities." *National Defense Magazine.* (http://bit.ly/1NsJvBK)

9) Wildlife Conservation unmanned aerial vehicle Challenge or wcUAVc. (http://bit.ly/1jr0pIE)

10) Ferdaus drone terror plot

Johnson, Kevin. "Man Accused of Plotting Done Attacks on Pentagon, Capitol." *USA Today.* (http://usat.ly/XXgiNw)

11) Fukushima radioactive drone protest

"Japanese Man Arrested for Landing Drone on PM's Office in Nuclear Protest." *Reuters*. (http://reut.rs/1TBTM3M)

12) NYC poster drone vandalism

Michel, Arthur. "The Age of Drone Vandalism Begins With an Epic NYC Tag." *WIRED.COM*. (http://bit.ly/1biAxuX)

13) Pirate Party Merkel drone stunt

"German 'Pirates' stage mini-drone stunt at Merkel rally." *RTE*. (http://bit.ly/1O6bqbs)

14) US troops withdrawn from Yemen

"Yemen Crisis: US Troops Withdraw from Air Base." *BBC News*. (http://bbc.in/1Is62CY)

15) Comparison of military strengths, US/Russia/China (http://bit.ly/1OJRT3j)

STAY CONNECTED

Visit the Swarm Troopers website for news and updates on the fast-moving world of drone technology.

THANK YOU

Thank you for reading Swarm Troopers, I hope you've found it an interesting and informative experience. If you enjoyed the book, please put up a short review on Amazon and help other readers to discover the world of Swarm Troopers.

ABOUT THE AUTHOR

David Hambling is a technology journalist and author based in South London. He writes for The Economist magazine, WIRED, Aviation Week, Popular Mechanics and Popular Science among others. His first book, Weapons Grade looked at the surprising military roots of modern technology.

Printed in Great Britain
by Amazon